Alveolar Structure and Function

Colloquium
Digital Library of Life Sciences

This e-book is a copyrighted work in the Colloquium Digital Library—an innovative collection of time saving references and tools for researchers and students who want to quickly get up to speed in a new area or fundamental biomedical/life sciences topic. Each PDF e-book in the collection is an in-depth overview of a fast-moving or fundamental area of research, authored by a prominent contributor to the field. We call these e-books *Lectures* because they are intended for a broad, diverse audience of life scientists, in the spirit of a plenary lecture delivered by a keynote speaker or visiting professor. Individual e-books are published as contributions to a particular thematic **series**, each covering a different subject area and managed by its own prestigious editor, who oversees topic and author selection as well as scientific review. Readers are invited to see highlights of fields other than their own, keep up with advances in various disciplines, and refresh their understanding of core concepts in cell & molecular biology.

For the full list of published and forthcoming Lectures, please visit the Colloquium homepage: www.morganclaypool.com/page/lifesci

Access to Colloquium Digital Library is available by institutional license. Please e-mail info@morganclaypool.com for more information.

Morgan & Claypool Life Sciences is a signatory to the STM Permission Guidelines. All figures used with permission.

Colloquium Series on Integrated Systems Physiology: From Molecule to Function to Disease

Editors

D. Neil Granger, *Louisiana State University Health Sciences Center*

Joey P. Granger, *University of Mississippi Medical Center*

Physiology is a scientific discipline devoted to understanding the functions of the body. It addresses function at multiple levels, including molecular, cellular, organ, and system. An appreciation of the processes that occur at each level is necessary to understand function in health and the dysfunction associated with disease. Homeostasis and integration are fundamental principles of physiology that account for the relative constancy of organ processes and bodily function even in the face of substantial environmental changes. This constancy results from integrative, cooperative interactions of chemical and electrical signaling processes within and between cells, organs and systems. This eBook series on the broad field of physiology covers the major organ systems from an integrative perspective that addresses the molecular and cellular processes that contribute to homeostasis. Material on pathophysiology is also included throughout the eBooks. The state-of the-art treatises were produced by leading experts in the field of physiology. Each eBook includes stand-alone information and is intended to be of value to students, scientists, and clinicians in the biomedical sciences. Since physiological concepts are an ever-changing work-in-progress, each contributor will have the opportunity to make periodic updates of the covered material.

Published titles

(for future titles please see the website, www.morganclaypool.com/page/lifesci)

Alveolar Structure and Function
D. Keith Payne and Adam Wellikoff
www.morganclaypool.com

ISBN: 9781615045044   paperback

ISBN: 9781615045051   ebook

DOI: 10.4199/C00067ED1V01Y201210ISP037

A Publication in the

COLLOQUIUM SERIES ON INTEGRATED SYSTEMS PHYSIOLOGY: FROM MOLECULE TO FUNCTION TO DISEASE

Lecture #37

Series Editors: D. Neil Granger, LSU Health Sciences Center, and Joey P. Granger, University of Mississippi Medical Center

Series ISSN

ISSN 2154-560X   print

ISSN 2154-5626   electronic

Alveolar Structure and Function

D. Keith Payne and Adam Wellikoff
Louisiana State University Health Sciences Center in Shreveport
Division of Pulmonary, Critical Care & Sleep Medicine

COLLOQUIUM SERIES ON INTEGRATED SYSTEMS PHYSIOLOGY:
FROM MOLECULE TO FUNCTION TO DISEASE #37

ABSTRACT

In the distal regions of the human lung, one of the most challenging problems facing a large multicellular organism is solved—ensuring an adequate supply of oxygen for aerobic tissue metabolism while removing associated waste products. Conduits for both air and blood converge at the alveolar level to match ventilation with perfusion and thus assure the free diffusion of oxygen and carbon dioxide. Despite their thin walls and their intimate relationship to the pulmonary capillary bed, the alveolus must present a barrier function robust enough to resist alveolar flooding from the hydrostatic pressures generated by the weight of the lungs and the volume of blood in the pulmonary circuit. The strategic position of the alveolar region and its vast associated capillary network ensure its importance in the synthesis and degradation of a wide range of molecules. Finally, the alveoli have evolved important immune functions vital to protecting the host from a variety of inhaled pollutants and microorganisms. Understanding alveolar structure and function is essential not only to appreciate the elegance of the human lung in its pristine state but also to understand the perturbations that underlay many lung diseases.

KEYWORDS

alveolus, terminal respiratory unit, alveolar macrophage, alveolar epithelium, alveolar capillary endothelium, lung development, surfactant, interstitial fibroblasts, pulmonary gas exchange, innate immunity, adaptive immunity, pulmonary host defenses

Contents

CHAPTER 1

Introduction

The spongy, pink surface of the normal human lung glides smoothly along the chest wall, expanding and contracting with each inhalation and exhalation in a cycle repeated thousands of times daily and requiring virtually no conscious thought. Over 10,000 liters per day of the ambient atmosphere (along with any associated dust, fumes, organic debris, microorganisms, etc.) are entrained through the mouth and nasopharynx, passing along approximately 17 generations of conducting airways before finally arriving at the level of the first alveoli. It is here that one of the major physiological problems facing a large multicellular organism is solved—ensuring an adequate supply of oxygen for aerobic tissue metabolism while removing the associated waste products. This is achieved in the lung by matching ventilation with perfusion at the alveolar level so that effective diffusion of oxygen and carbon dioxide may occur. Due to the low solubility of oxygen in water, the alveolar walls must be quite thin, and there must be a lot of them—around 500 million alveoli (total, both lungs) in an average sized human [1]. Although considerable variability exists depending on which part of the lung is measured, an "average" alveolus has a diameter of about 250 μm with an average of 1000 capillaries entwined around it. With 85–95% of the alveolar surface surrounded by capillaries, the effective diffusion membrane created (total, both lungs) approaches 80 m^2, not quite as large as the singles half of a tennis court [2]. While diffusion of oxygen and carbon dioxide is clearly the major role of the alveoli, they have other important functions as well. Despite their thin walls and their intimate relationship to the pulmonary capillary bed, the alveolus must present a barrier function robust enough to resist alveolar flooding from the hydrostatic pressures generated by the weight of the lungs and the volume of blood in the pulmonary circuit. The strategic position of the alveolar region and its vast associated capillary network ensure its importance in the synthesis and degradation of a wide range of molecules. Finally, the alveoli have evolved important immune functions vital to protecting the host from a variety of inhaled pollutants and microorganisms. A clear understanding of the importance of the structure and function of the alveolar region of the lung is essential not only to appreciate the elegance of the lung in its pristine state but also to understand the perturbations that underlay many lung diseases.

. . . .

CHAPTER 2

Historical Perspectives

The history of the lungs and the mechanics of gas exchange as achieved in the alveoli owe much to the discovery and observation of the pulmonary circulation. Thanks to the work of people like Hippocrates around 400 BC, Galen in the 1st century AD, Michael Servetus in the 16th century and William Harvey in the 17th century, the course of blood through the heart and pulmonary circulation was defined in great detail. Their work up to the end of the 17th century described how blood moved from the body to the right side of the heart and, as William Harvey put it in his book "De Motu Cordis" in 1628, from the right heart chamber through the *vena arteriosa* (pulmonary artery) through the "substance" of the lungs where it "mingles" with air [3]. How this "mingling" took place was still somewhat of a mystery.

Not until 1661 did Marcello Malpighi, using a light microscope, identify not only the capillaries but also the alveoli as well. In a letter to Giovanni Borelli, Malpighi describes, ". . . the whole mass of the lungs, with the vessels going out of it attached, to be an aggregate of very light and very thin membranes, which, tense and sinuous, form an almost infinite number of orbicular vesicles and cavities, such as we see in the honeycomb alveoli of bees, formed of wax spread out into partitions" [4]. This is the first notation of the word "alveoli" used to describe what would later be discovered to be the gas exchange unit of the lung. Malpighi went on to say that, "These [vesicles and cavities] have situation and connection as if there is an entrance into them from the trachea, directly from the one into the other; and at last they end in the containing membrane" [4]. Included in the description of the alveoli is the observation that the membrane that makes up these "vesicles" is "endowed" with the vessels communicating with the larger arteries. He observed that the vessels leading into the lung distribute themselves into the membranes of the alveoli at such a small scale that they ". . . escape the senses on account of their exquisite smallness" [4]. The vessels then reappear and coalesce to form the pulmonary vein leading to the left ventricle. Malpighi went a step further and was the first to show that the membrane that separates the vessel from the part of the vesicle that communicates with the trachea is permeable. He poured black water and mercury into the pulmonary artery of a dissected lung and noted that these substances penetrated into the smallest of vessels. Then, when slightly compressing the lung noted the water and mercury to "sweat out" from the membrane, partially collect in the interstitium, and come out of both the pulmonary vein and the trachea. It was noted, however that the water coming out of the trachea was slightly frothy and of less color. This hinted at the selective permeability of the alveolar membrane.

Thanks to the work of these early pioneers, the circulation of blood through the lungs was more clearly defined. What this circulation accomplished and how it contributed to life remained elusive for more than a century. Throughout the 17th and 18th centuries, people like Robert Hooke, Carl Wilheim Scheel, Joseph Priestly, Antoine Lavoisier and Joseph Black helped to discover the presence of oxygen and carbon dioxide in the air. It was Sir Humphry Davy, however, at the end of the 18th century who documented the presence of these molecules in the blood. He did this by heating the blood and collecting the gas that was produced. Davy went one step further to say that oxygen in the blood was carried to the tissues where energy was released in the form of carbon dioxide that was then carried away [5, 6]. Gustav Magnus then measured the amount of these gases in the blood and found more oxygen than carbon dioxide and demonstrated that this exchange took place in the lungs [7, 8, 9]. A controversy raged for the next half century between whether oxygen exchange between alveolar gas and pulmonary capillary blood occurred via active secretion or by simple diffusion.

On the side of oxygen secretion were the likes of Christian Bohr (1855–1911) and J.S. Haldane (1860–1936) citing the example that "... normal oxygen tension in the arterial blood is always higher than the alveolar air" [10, 11]. This difference in oxygen tension on either side of the alveolar membrane, or blood–gas barrier, did not allow for diffusion as this principle is based on molecules traveling from an area of high concentration to that of a lower one. What Bohr and Haldane did not realize, however, was that the technique they used to measure oxygen tension in the blood was inaccurate. August Krogh (1874–1949) along with his wife, Marie (1874–1943) used a more accurate tonometer to measure arterial oxygen content. They found that the oxygen tension of arterial blood was, in fact, always lower than that of alveolar air. This was shown in a variety of animal experiments and settled the argument definitively in favor of diffusion.

. . . .

CHAPTER 3

Lung Development

Our lungs are unique compared to the other organ systems of the body. Other than the GI tract and the skin, it is the only system exposed to the outside environment. The GI system and skin, however, along with the other organ systems start functioning early in fetal life and progressively increase their capabilities throughout gestation. The lungs are filled with amniotic fluid while gestating and gas exchange takes place in the mother's lungs with oxygenated blood passing to the developing fetus via the umbilical vein. At the instant of birth, with the separation of the newborn baby from the placenta, the lungs have to assume the role of gas exchange so that the baby can begin his or her life.

In order for the lung to be able to provide respiration and gas exchange at birth, it must go through multiple morphologic changes. The development of the respiratory system is a continuous process with alveolar development occurring quite late in the process and even continuing through the postnatal period. There are multiple developmental stages that occur based on both anatomic and histologic criteria (see Figure 3.1). Although the focus of this chapter will be on the development of the pulmonary acinus, that is, the gas exchanging units of the lung or terminal respiratory unit (respiratory bronchioles and alveoli), we will briefly cover the other stages. If not for these stages, there would be no pathway to the acinus and therefore no gas exchange.

3.1   EMBRYOLOGY AND PRENATAL DEVELOPMENT

The lung first begins to form around the third week of gestation. This early phase, called the embryonic phase, is characterized by the formation of a ventral out-pouching from the laryngotracheal groove on the foregut endoderm [12, 13]. The endoderm is one of the three cell types seen in the early embryo and is the inner most layer. The other two are the ectoderm, the outer most layer and the mesoderm, the layer in the middle. The laryngotracheal groove separates the esophagus from the trachea and eventually gives rise to two primary bronchial buds (see Figure 3.2). These bronchial buds, composed of endodermal cells that become the epithelium, grow through the surrounding mesoderm that will eventually become an undifferentiated collection of loose connective tissue called the mesenchyme. This epithelio-mesenchymal interaction promotes monopodial (from Greek "mono-," one and "-podial," foot) lateral branching at first, but eventually leads to dichot-

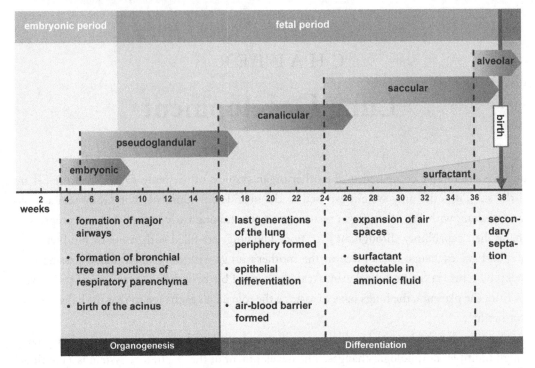

FIGURE 3.1: Phases of lung development. (Reproduced with permission from *Phases of lung development*. Module 18: Respiration tract [cited 2012 March 18]; Available from: http://www.embryology .ch/anglais/rrespiratory/phasen01.html.)

Stage 1—Embryonic period (3–7 weeks)
    —Lung budding from the foregut endoderm
    —Formation of trachea and mainstem bronchi
Stage 2—Pseudoglandular phase (5–17 weeks)
    —Airway division completed with 25,000 terminal bronchioles
    —Cartilage and smooth muscle derived from the mesenchyme
Stage 3—Canalicular phase (16–24 weeks)
    —Capillarization — formation of the blood–gas barrier
    —Acinar formation, types I and II pneumocytes first differentiate
Stage 4—Saccular phase (24–36 weeks)
    —Progressive thinning of epithelial cells
    —Terminal saccular formation
    —Surfactant production
Stage 5—Alveolar Phase (36 weeks through infancy)
    —Appearance of true alveoli
    —Alveolar septation and expansion of air spaces

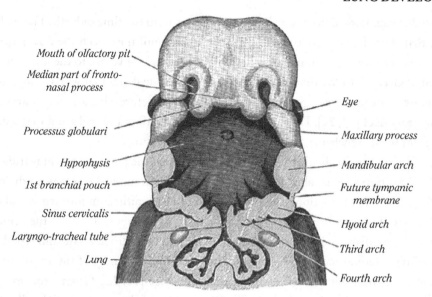

Mouth of olfactory pit
Median part of fronto-nasal process
Eye
Processus globulari
Maxillary process
Hypophysis
Mandibular arch
1st branchial pouch
Future tympanic membrane
Sinus cervicalis
Hyoid arch
Laryngo-tracheal tube
Third arch
Lung
Fourth arch

FIGURE 3.2: Laryngotracheal groove. Schematic representation of the appearance of the laryngotracheal groove *in utero*. (Reproduction of a lithograph plate from Gray's Anatomy from the 20th U.S. Edition originally published in 1918 and has therefore lapsed into the public domain.)

omic (meaning equal division of a terminal bud) branching [14, 15]. This process, which eventually leads to not just two lungs, but two lobes on the left, three lobes on the right and numerous distal subdivisions, is controlled by the complex interaction of the mesenchymal and epithelial layers. At this early stage, the lung epithelium consists of tall columnar cells and shows no evidence of morphologic differentiation. There is, however, evidence of molecular differentiation [16]. The different cell layers (endoderm or epithelium, mesoderm or mesenchyme) and the basal lamina that separate them contain a multitude of molecules including surfactant protein C (SP-C), fibroblast growth factor (FGF), transforming growth factor-β-1 (TGF-β-1) and even one named Sonic Hedgehog (Shh) as well as many others [17, 18]. The mesenchymal cells that surround the developing lungs will give rise to blood vessels, smooth muscle, cartilage and other connective tissue, whereas the epithelium will become a barrier to the outside environment and the gas exchanging units [12, 19]. All of these developmental milestones occur during the embryonic phase of development, which is only the beginning.

During the next phase of development, the pseudoglandular phase (5–17 weeks), the early tubular system, that is, the airways, is lined with pseudostratified columnar epithelium. The distal parts of this tubular system instead of being lined by pseudostratified epithelium are lined by cuboidal cells that differentiate into type II pneumocytes. These bronchopulmonary epithelial cells begin to produce amniotic fluid that will fill up the lungs until the time of birth. The dichotomous branching that started earlier continues during this phase until they reach the final pattern of the pulmonary tree resulting in terminal bronchioles that eventually give rise to the acinar tubules that form the adult alveoli [16]. By the end of this phase, the terminal sacs that will develop into the gas exchange units of the lung are beginning to form as dilated structures that arise in clusters at the end of the respiratory duct [20, 21]. During this period, the smooth muscle cells and vascular branches that form from the mesenchyme do so parallel to the epithelium.

It is not until the canalicular phase (16–24 weeks) that the pulmonary acinus truly begins to develop. As the airways get smaller, the mesenchyme that surrounds the airway epithelium thins, allowing for a close approximation of the capillaries with the epithelium forming vascular "canals" that will become the areas for gas exchange in the lungs. The cuboidal cells of the acinar tubules also begin to differentiate into type II pneumocytes that, in turn, differentiate into type I pneumocytes. This differentiation into type I pneumocytes leads to a flattening of the alveoli making this future blood–gas barrier thin enough to support gas exchange [18, 22]. Other structures that form during this period include submucosal glands, Clara cells and the cartilage of the distal terminal bronchioles. Lamellar bodies, which are located inside the type II pneumocytes and are involved in surfactant production, also become visible [12].

Once the developing fetus enters the third trimester, the terminal acinar tubules continue to branch and increase the size of the air spaces leading to saccule formation, hence the term saccular phase (24–36 weeks). Thin type I pneumocytes start to outnumber the type II pneumocytes thereby increasing the area available for gas exchange. This marks the threshold for viability of the developing fetus and preterm infants born during this time may survive if they have access to neonatal intensive care units [23, 24]. How this branching occurs is a matter of on-going research and discussion. One theory involves a complex mechanism of inhibitor and promoter molecules. FGF-10 is responsible for bronchial bud and epithelial proliferation, whereas Shh and TGF-β-1 inhibit it. TGF-β-1 is also responsible for secretion of several extracellular components, including elastin, to provide support for the growing saccules [21]. The other theory involves expansion of the terminal sacs by increased intraluminal pressure in conjunction with molecular signaling. We have already discussed that the epithelial cells in the developing lung produce some of the amniotic fluid. This fluid production results in an increase in the pressure inside the lumen of the developing airways. As this intraluminal pressure increases, the terminal buds balloon through areas of decreased resistance and are inhibited when they encounter areas of higher resistance. This theory still involves some signaling molecules, such as FGF-10, that produce the extracellular components that likely account for the differences in resistance encountered by the expanding terminal sacs [21].

Terminal saccules become true alveoli during the last stage of development, the alveolar phase, which continues from the 36th week of gestation to several months into the post-natal life.

Much like an orchestra must work together to produce beautiful music, several cell types must work together to form the alveoli. This stage is characterized by the formation of septa in the walls of the terminal saccules in order to increase the surface area available for gas exchange. A good way to picture this is to imagine the walls of the terminal saccules as the walls of an empty room. The septa then form like shelves being placed onto the walls of the room. The shelves only protrude part of the way into the room, much like the septa protrude into the saccules. These secondary septae grow into the room, but usually stop short of connecting to each other leading to the formation of the alveolar ducts. It is the myofibroblasts that produce elastin in the mesenchyme that must work together with the airway epithelium and the vascular endothelial cells to allow this "alveolarization" to occur [21]. Elastin is the linchpin in this process as septa form at the location of elastin deposition. This is supported by evidence that mice bred with an elastin knockout gene will lose septa and develop emphysema [25]. Exactly how the myofibroblasts deposit elastin at the locations that they do is a matter of on-going research. Suffice to say that this process likely involves a close interaction between signaling molecules, dormant stem cells, and proximity to myofibroblasts, alveolar epithelial cells and vascular endothelium.

In order for final, mature alveoli to form and thereby support gas exchange, two more processes must occur. These are the formation of a mature microvasculature and further thinning of the septal mesenchyme. The thinning of the mesenchyme is vital as gas exchange can only occur through a very thin blood–gas barrier. Derangements in this barrier that increase its thickness result in poor gas exchange as seen in conditions such as congestive heart failure and idiopathic pulmonary fibrosis. This thinning occurs as a result of apoptosis and occurs simultaneously with ongoing expansion of the epithelial and blood vessel components, allowing for a close approximation of the airspaces with the vasculature [26, 21]. As with many other stages of fetal lung development, the exact mechanism by which the formation of a mature microvascular circulation forms is not fully known. It is known, however, that multiple signaling molecules, such as vascular endothelial growth factor (VEGF), platelet-derived growth factor-B (PDGF-B) and angiopoeitin-1 and -2 are involved.

3.2   POSTNATAL DEVELOPMENT

The alveolarization that occurs prior to birth only results in about 30 million of the roughly 300–500 million fully mature alveoli present in the adult lungs. The alveoli that have developed, however, are not fully formed [25, 27]. The septa that began to form during the alveolar phase of prenatal development are composed of two capillary layers separated by a central sheet of connective tissue. This is in contrast to the fully mature septa that only include one capillary layer and its supporting connective tissue [27] (see Figure 3.3). It is this morphologic appearance that separates immature saccules from mature alveoli as these immature capillaries face the airspace on only one side with

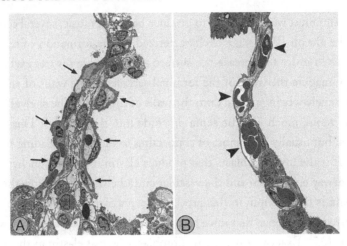

FIGURE 3.3: Comparison of alveolar septa. Comparison of interalveolar walls in a human lung at 26 days (A) and in an adult lung (B) at a magnification of 1540×. (Reproduced with permission from Burri, P.H., *Structural aspects of postnatal lung development—alveolar formation and growth. Biol Neonate*, 2006. 89(4): p. 313–22.)

(A) The septum is immature with a central sheet of interstitial tissue (it) flanked on both sides by a capillary layer. The arrows point to the blood–gas barrier.

(B) In the adult lung a single capillary network meanders through the septum (arrowheads) crossing the layer of supporting connective tissue (dashed line).

the other side covered by a thick sheet of connective tissue. These septa are called "primary septa" and form the basic structures needed for final alveolarization [28].

The term "bulk alveolarization" has been used to describe the rapid changes seen in the lungs in the first few months after birth. This period is characterized by the formation of secondary septa that begin to form off primary septa and lead to an increase in the number of alveoli [27] (see Figures 3.4A and 3.4B). This process is mediated by elastin deposition just as it is in the alveolar phase of prenatal development with the walls of these septa composed of two capillary layers with a central sheet of connective tissue. It is not until the baby is around 3 years old that the process of septal restructuring is complete. This process involves apoptosis of the septal interstitium allowing for progressively closer approximation of the two capillary walls until they eventually merge.

Although the lung is filled with amniotic fluid, the structure of the alveoli does not change at birth. During birth, much of the fluid in the respiratory tree is expelled as the chest squeezes through the narrow birth canal. The remaining fluid is removed by the pulmonary circulation and lymphatic system after birth [29]. The alveolar spaces that have developed at the time of birth, once

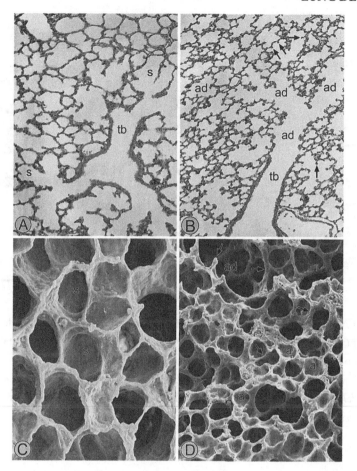

FIGURE 3.4A: Formation for secondary septa after birth. Formation of the alveoli in the rat lung documented by light microscopy at 32× magnification (A, B) and by scanning electron microscopy at 460× magnification (C, D). (Reproduced with permission from Burri, P.H., *Structural aspects of postnatal lung development—alveolar formation and growth. Biol Neonate*, 2006. 89(4): p. 313–22.)

(A) Day 1: Terminal bronchiole (tb) opens into a smooth-walled channel dividing into several saccules (s).

(B) Day 21: The terminal bronchiole (tb) now opens into several generations of alveolar ducts (ad) surrounded by alveoli (arrows).

(C) Day 1: Lung parenchyma made of smoothly-lined saccules (s).

(D) Day 21: By the formation of the secondary septa (arrows) the smooth-walled channels and saccules of (A) have been transformed into alveolar ducts (ad) and alveolar sacs (a), both lined with alveoli.

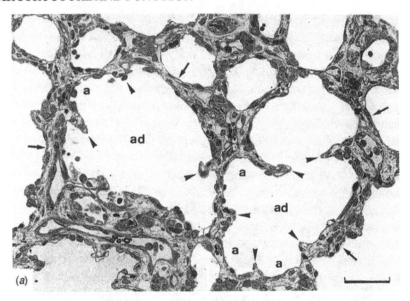

FIGURE 3.4B: EM of secondary septae formation. Light micrograph of human lung aged 26 days. Thick primary septa (arrows) and new secondary septa (arrowheads). All septa contain at least part of a double capillary network. a = alveoli, ad = alveolar duct. Magnification: 250×. Bar = 50 μm (Reproduced with permission form Burri, P.H., *Structural Development of Lung in Fetus and Neonate, in Fetus and Neonate: Physiology and Clinical Applications: Volume 2, Breathing*, M. Hanson, et al., Editors. 1994, Cambridge University Press: Great Britain. p. 3–19.)

emptied of the amniotic fluid, contain residual surfactant produced by the type II pneumocytes. When the first breath of air enters these spaces, surfactant disperses and forms stable bubbles resulting in the formation of surface tension. High expiratory pressure then further distributes the surfactant [30]. The alveoli are distended by these bubbles as well as by inspired air. This establishes the functional residual capacity that will persist throughout the respiratory cycle.

·   ·   ·   ·   ·

CHAPTER 4

Structural Anatomy—Conducting Airways to the Alveoli

There are normally 23 generations of continuously branching airways, beginning with the trachea and ending in the region of the alveolar ducts. Significant gas exchange only takes place in the last 7 generations of airways, the first 16 being conducting airways, devoid of alveoli (see Figure 4.1). The average human trachea is around 22–26 mm in diameter. The airways progressively narrow in size down to a width of 0.5 mm or even less in the region where alveoli first begin to appear in the walls of the respiratory bronchioles. The lung is designed so that the total cross-sectional area of the smaller airways greatly exceeds that of the larger ones, thus making gas exchange possible over a very large surface area. Branches of the pulmonary artery along with lymphatics and nerve fibers follow the conducting airways closely down to the level of the respiratory bronchioles (see Figure 4.2). All of these structures are supported by a stroma of connective tissue composed of a matrix of gel-like proteoglycans, a variety of interstitial cells and collagen, elastin and reticulin fibers. Collectively, this is referred to as the extra-alveolar interstitium of the lung. A similar compartment of this matrix, the parenchymal (alveolar) interstitium, forms an elastic structure around the millions of tiny alveoli at the very end of the airways. The alveoli are intimately connected with the wall of one alveolus comprising the wall of its neighbor (see Figure 4.3). Both the alveolar and extra-alveolar interstitium are linked together as a continuous band of supportive tissue, anchored distally by attachments along the surface limiting membrane of the lung, the visceral pleura and anchored centrally along the fibrous connections to the conducting airways. Because of this, tethering effect of the total interstitial component of the lung the alveoli maintain their shape throughout all stages of normal lung inflation and deflation, flattening out only at minimal lung volume (see Figure 4.4a and Figure 4.4b). Once this occurs, the role of surfactant (covered later in this book) becomes particularly important in reducing the surface tension of compressed, flattened alveoli and allowing their reinflation with a minimum of effort. Additional structures that may contribute to alveolar stability, particularly in the face of localized small airway obstruction, include the pores of Kohn and the canals of Lambert. The pores of Kohn are small holes 5–15 μm in diameter in the walls of

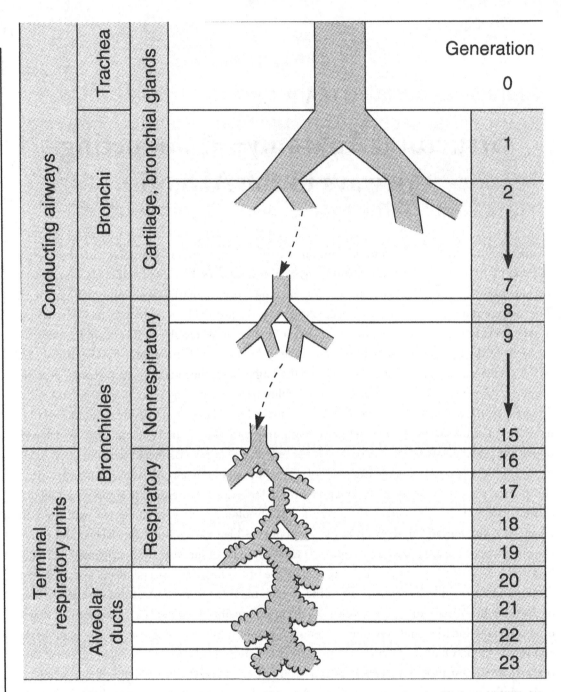

FIGURE 4.1: Subdivisions of the human airways from the trachea to the alveoli. Generation 0 (the trachea) down to generation 15 are conducting airways. Gas exchange takes place in the respiratory airways from generations 16 to 23, the region where alveoli are located. The cross-sectional area of the smaller airways greatly exceeds that of the larger airways, making gas exchange more efficient. (Reproduced with permission from McPhee SJ and Ganong WF: Pathophysiology of Disease: An Introduction to Clinical Medicine, 5th Edition, p. 219, 2006, The McGraw-Hill Companies, Inc.)

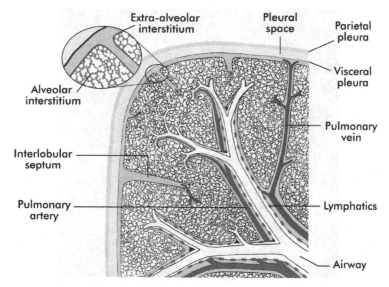

FIGURE 4.2: General organizational structure of the lung. Note the lymphatics (green), and vascular structures including the pulmonary artery (blue) and bronchial artery (not depicted here), follow the conducting airways. The interstitial compartments of the lung consist of the connective tissue surrounding the alveoli (the alveolar interstitium), which is connected in turn to the extra-alveolar interstitium. The latter is connected distally to the visceral pleura and the interlobular septae, and centrally to connective tissue along the airways. The pulmonary veins (red) drain through the interstitium of the lung away from the airways until they reach the hilum of the lung on their way to the left atrium. (Modified and reproduced with permission from Hayek, H: The Human Lung. New York: Hafner, 1960, p. 52.)

adjacent alveoli that permit the passage of air, cells, and bacteria (see Figure 4.5). They probably vary in diameter during inflation and deflation cycles and may well function to provide collateral ventilation between adjacent alveoli, although there is some controversy about this role [31, 32]. The canals of Lambert are openings in the walls of terminal and respiratory bronchioles that communicate directly with alveoli.

4.1   THE TERMINAL RESPIRATORY UNIT

The terminal respiratory unit (TRU) was first described by the German anatomist von Hayek and consists of the respiratory bronchiole and all its associated alveolar ducts and alveoli (see Figure 4.6). Some authors also refer to the TRU as the acinus or primary pulmonary lobule. In human lungs, a single terminal respiratory unit consists of about 100 alveolar ducts and 2000 alveoli. The second-

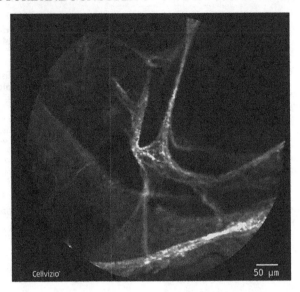

FIGURE 4.3: http://tinyurl.com/video4-3. Video of human alveoli in vivo photographed with the use of probe-based confocal laser endomicroscopy. Thin band-like fibers are the elastin network that makes up the support structure of the alveoli. Larger, solid longitudinal structure on the lower left is a small blood vessel.

ary pulmonary lobule consists of 3–5 TRUs. Secondary pulmonary lobules are separated from each other by extra-alveolar interstitium known as the interlobular septa. These septa contain lymphatic channels which may be visible on chest x-rays as "Kerley's B lines" when they are filled with fluid, inflammatory cells or scar tissue. The lining of the conducting airways consists of ciliated columnar epithelium down to the level of the respiratory bronchiole where there is an abrupt transition to the very thin and delicate alveolar epithelium. The inside lining of each alveolus is composed of two distinct cell types conveniently named type I and type II alveolar epithelial cells (sometimes called pneumocytes). They appear to be very different in both structure and function as discussed later. Type I cells are flat in shape and less numerous than the more cuboidal type II cells, yet type I cells cover 90–95% of the inside alveolar walls. They appear to have more of a barrier and gas exchange function compared to the type II cells, which are primarily involved in alveolar epithelial repair and synthetic activities.

4.2 INNERVATION

The human lung is innervated by branches of the thoracic sympathetic ganglia and the vagus nerve (parasympathetic pathway). These nerve branches enter the lung at the hilar regions and accompany the bronchi, pulmonary arteries and pulmonary veins to invest the airways, smooth muscle and

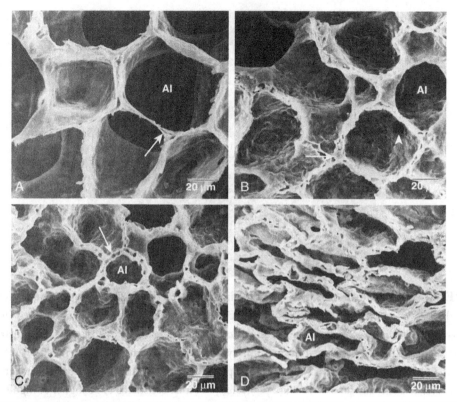

FIGURE 4.4A: Alveolar shape at various stages of distension. Alveolar shape changes from:
(A) Total Lung Capacity (TLC)—at 30 cm H_2O of pressure
(B) 50% TLC—about 8 cm H_2O of pressure
(C) Functional Residual Capacity (FRC)—at 4 cm H_2O of pressure
(D) Residual Volume (RV)—at 0 cm H_2O of pressure.

Vascular pressures are held constant throughout. Overall alveolar shape (AI) is maintained from TLC to FRC. Note how the alveolar capillaries (arrows) are slit-like at TLC and are rounded at FRC. Pulmonary vascular resistance increases at high lung volumes likely due to the narrowing of the alveolar capillaries. The arrow in (B) identifies a type 2 pneumocyte at the corner of an alveolus. (Perfusion-fixed normal rat lungs, scanning electron microscopy) (Reproduced with permission from Murray and Nadel's Textbook of Respiratory Medicine, 5th edition, 2010, p. 14, Philadelphia, Saunders Elsevier.)

vascular structures. Both efferent and afferent fibers are carried in the vagus nerve. Afferent sensory nerve fibers have been traced a far as the respiratory bronchioles [33]. Unmyelinated efferent nerve fibers appear to extend to the regions of the alveolar ducts where they innervate bands of smooth muscle [34]. Another group of unmyelinated axons, C-fibers (sometimes called J receptors) are also found in the terminal respiratory units where they may be located in the interstitial areas around

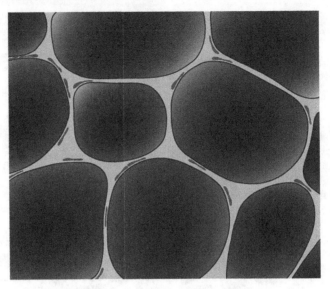

FIGURE 4.4B: http://tinyurl.com/video4-4b. Animation depicting alveolar shape changes (exaggerated for the sake of clarity) occurring during the respiratory cycle. As previously mentioned in Figure 4.3a, alveolar shape changes are maintained down to minimal lung volumes and alveolar capillaries also change shape. Because of a greater transpulmonary pressure gradient (the difference between pleural pressure and alveolar pressure) alveoli in the lung apex are larger and the capillaries are smaller than in the lung bases. The slit-like capillaries cause pulmonary vascular resistance to be higher at high lung volumes. This phenomenon along with the effect of gravity on pulmonary blood flow contributes to regional variations in ventilation/perfusion ratios between the uppermost and lower lung zones (see graphical depiction of lung zones in Figure 4.11).

respiratory bronchioles or in the alveolar walls themselves. They may play a role in sensing interstitial distortion from edema caused by pulmonary vascular congestion or inflammation [35]. The distal airways out to the region of the alveolar ducts also contain components of the neuroendocrine system. Neuroepithelial bodies are frequently located in the bifurcation of very small distal airways. They appear to be storage sites for substance P and vasoactive intestinal peptide [36, 37]. Although their function is poorly understood, it is possible they act as peripheral chemoreceptors. In animal models, acute hypoxia or hypercapnia causes neuroendocrine cells to degranulate [38].

4.3   ALVEOLAR CELLULAR STRUCTURE

Cells in the alveolar region of the lung include the type 1 and type 2 alveolar epithelial cells which together form the interior wall of the alveolus, the capillary endothelial cell which for practical purposes composes the majority of the exterior wall of the alveolus, and interstitial cells such as

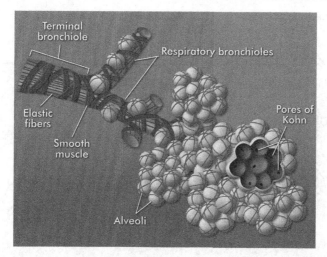

FIGURE 4.5: The distal airways. Structure of the distal portion of the airways from the terminal bronchiole to the alveoli. Note the appearance of the first alveoli at the level of the respiratory bronchioles. Elastic fibers (green) surrounding the alveoli are part of the alveolar interstitium and provide crucial tethering support during inflation and deflation cycles of the lung. The diameter of the terminal bronchiole is about 1 mm although this varies by the size of the individual and inflation state of the lung.

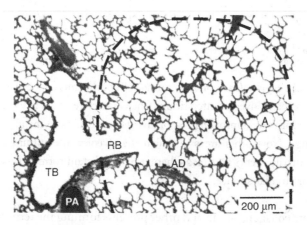

FIGURE 4.6: Light microscopy of the terminal respiratory unit. Light microscopic image of the terminal respiratory unit (dashed line) from normal sheep lung. "A" is an alveolus. TB is a terminal bronchiole. RB is a respiratory bronchiole. AD is an alveolar duct. PA is a pulmonary artery (note close approximation to the airway). PV is a pulmonary vein (note peripheral location). (Reproduced with permission from Murray and Nadel's Textbook of Respiratory Medicine, 5th edition, 2010, p. 10, Philadelphia, Saunders Elsevier.)

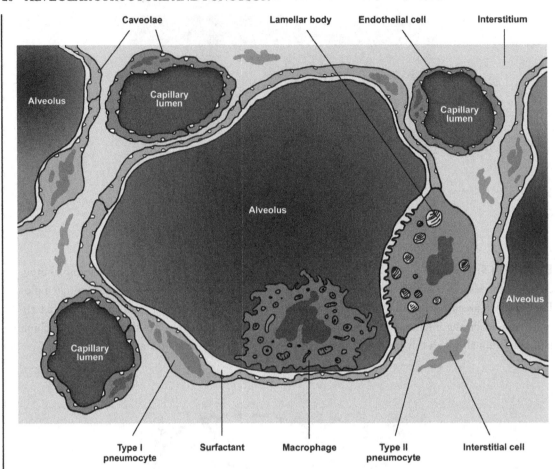

FIGURE 4.7: The alveolar region. Cartoon illustrating cells of the immediate alveolar region of the lung, including the alveolar walls and alveolar interstitium (the connective tissue between alveoli including interstitial cells). Note the intimate approximation of alveolar capillaries to the alveolar walls, minimizing distance for gas exchange. Note other areas in which the alveolar interstitium is much thicker—fluid and solute exchange primarily occurs here (see Chapter 6). Surfactant forms a very thin continuous layer across the entire interior alveolar surface.

fibroblasts which secrete collagen, elastin and other proteins forming the tethering supporting structure of the small airways and alveolar region and the alveolar macrophage (see Figure 4.7). Table 4.1 lists the approximate number and percentage of cells in the alveolar region of the human lung.

4.3.1   Type I Alveolar Epithelial Cells

Compared to other cells of the alveolar region, relatively little is known about the type I alveolar epithelial cell. It has proven to be difficult to isolate and study *in vitro* and much of the available data

CELLS	$n \times 10^9$	TOTAL (%)
TABLE 4.1: Cells of the alveolar region of the human lung.		
Alveolar epithelial cells		
Type I cells	19	8.3
Type II cells	37	15.9
Endothelial cells	68	30.2
Interstitial cells	84	36.1
Macrophages	23	9.4

From Lippincott Williams and Wilkins, Chest Medicine, *Essentials of Pulmonary and Critical Care Medicine*, 5th ed 2005. Original permission from Crapo JD, Barry BE, Gehr P, et al. Cell number and characteristics of the normal human lung. *Am Rev Respir Dis* 1982; 125: 740–745.

come from animal, not human, studies. Type I cells are extremely flat, a property that is thought to be advantageous in promoting gas exchange across the alveolar/capillary membrane. They account for around 8% of the total cell population in the lung yet constitute 95% of the alveolar surface area [39]. Type 1 cells may extend their thin cytoplasmic processes into adjacent alveoli. They appear to be susceptible to injury from a variety of insults including inhaled toxins, oxidant stresses, barotrauma, etc. compared to other cell types. When this occurs, they are replaced by type II alveolar epithelial cells which lose their normally cuboidal shape as well as their cell marker genes such as SP-A, SP-B and SP-C and transform into the flattened type 1 cell phenotype, complete with type 1 cell gene expression. It has previously been thought that type 1 cells were terminally differentiated and unable to undergo cell division. Recent data from isolated rat type 1 cells indicate that indeed they may undergo cell division, at least *in vitro* [40]. Confirmation of this is needed *in vivo*. Type I cells contain many vesicles, or calveolae, which open into the alveolus or the interstitial space. These vesicles contain a protein, caveolin-1, which is thought to play an important role in sequestration of proteins into areas of the cell membrane and in receptor inactivation. The overall effect of this may be to assist in maintaining alveolar homeostasis by keeping sequestered proteins and receptors in a quiescent state.

Type I cells produce a limited number of proteins compared to type 2 cells (see Table 4.2. http://tinyurl.com/alveolar4-2). Aquaporin-5 (AQP-5) is a water channel that has been shown to convey remarkable water permeability. This finding along with the existence of enzymes such as

TABLE 4.2: Products synthesized by types I and II alveolar epithelial cells.

Receptors		Enzymes		Channels, Transporters, Miscellaneous		Surfactant Associated Proteins		Matrix Proteins		Growth Factors & Cytokines	
Type I	Type II	Type I	Type II	Type I	Type II	Type I	Type II	Type I	Type II	Type I	Type II
IGFR-2	c-met	Na⁺/K⁺-ATPase	Na⁺/K⁺-ATPase	Aquaporin-5	Aquaporin-1	None	SP-A	None	Laminin, fibronectin	None	PTHrP
RAGE	RAGE	γ-Glutamyl transferase	γ-Glutamyl transferase	Aquaporin-4	Aquaporin-3		SP-B		Basement membrane components		Polypeptide-1
Purinergic receptor 4 (P2X4)	P2-purinergic receptor	Lipid phosphatidic acid phosphatase	Collagenase	Connexins 43, 46	Connexins 26, 32, 40, 43		SP-C		Tenascin		Monocyte chemotactic
B₂-adrenergic receptors	B-adrenergic receptor	CP-M	Carbonic anhydrase II	Plasminogen activator inhibitor 1	Plasminogen activator inhibitor		SP-D		Type IV collagen		Endothelin
	Cannabinoid receptor	Alkaline phosphatase	Alkaline phosphatase	gp 60	CFTR				Entactin		IL-1
	HDL receptors		Aminopeptidase N	VAMP-2	Adducin						IL-8
	SP-A, SP-B, SP-C		Gelatinases A, B	P15^{Ink4B}	α 1-antitrypsin						VEGF
	Toll-like receptor 2		Tripeptidyl tripeptidase I	Eotaxin	Cysteine transporter						CGRP
	EGF-R		α-glucosidase	P-glycoprotein	eNOS, iNOS						EGF
	ANP-R		Superoxide dismutase	eNaC	K⁺ channels						PDGF B chain
	FGF-R		Lamellar body lysozyme	ICAM-1	Na⁺ channels						
	TGF β-R		Cytochrome P-450 enzymes	Caveolin-1	Annexins						
	IFN-γ-R		Plasminogen activator	T1α	Surfactant phospholipids						
	PDGF-R		CD 44		Leukotriene C4						
	PTHrP-R		LAR-PTP		α1-acid glycoprotein						
	GM-CSF-R		PKC6		5-lipoxygenase activating protein						
	IL-2-R		CTP		Muc 1						
	Type II MHC				CC10						
	gp 330 (megalin)				Tissue factor						
					Pneumocin						
					Lipocortin 1						
					Maclura pomifera binding proteins						
					ABC transporter ABCA3						
					Cl⁻/HCO₃⁻ exchanger AE2						

ANP, atrial natriuretic peptide; CC10, clara cell secretory protein: uteroglobulin; CFTR, cystic fibrosis transmembrane regulator: CGRP, calcitonin gene–related peptide; CP-M, carboxypeptidase M; CTP, phosphocholine cytidyltransferase; EGF, epidermal growth factor; eNaC, epithelial sodium channel; eNOS, endothelial nitric oxide synthase; FGF, fibroblast growth factor; gp, glycoprotein; GM-CSF, granulocyte-macrophage colony-stimulating factor; HDL, high-density lipoprotein; ICAM -1, intracellular adhesion molecule-1; IFN-γ, interferon–γ; IGFR-2, insulin-like growth factor receptor-2; IL, interleukin; INK4B, cyclin-dependent kinase 4 inhibitor 2B; iNOS, inducible nitric oxide synthase; LAR-PTP, leukocyte common antigen–protein tyrosine phosphatase; MHC, major histocompatibility complex; PDGF, platelet-derived growth factor; PKC6, protein kinase Cδ; R, receptor; RAGE, receptor for advanced glycation end products; SP, surfactant protein; TGF-β, transforming growth factor–β; VAMP-2, vesicle-associated membrane protein-2; VEGF, vascular endothelial growth factor. For full-sized, horizontally aligned table: http://tinyurl.com/alveolar4-2

For full-sized, horizontally aligned table: http://tinyurl.com/alveolar4-2

Na^+/K^+-ATPase in type 1 cell membranes and epithelial sodium channels may indicate that type 1 alveolar epithelial cells play an important role in fluid and solute flux across the alveolar/capillary membrane.

4.3.2   Type II Alveolar Epithelial Cells

Type 2 alveolar epithelial cells make up only 5% of the total alveolar surface. Unlike the type 1 cells, they are cuboidal in shape with short microvilli on their luminal surface. They are typically located near the alveolar wall corners where they tightly attach to type 1 cells. They appear to have a cell cycle of 28–35 days although this may accelerate after lung injury [41, 42]. Much more is known about type 2 cells than their type 1 counterparts. Type 2 cells appear to have at least 4 important roles in the alveolus including synthetic capabilities, fluid and ion transport across the alveolar/capillary membrane (Chapter 6), replacement of damaged or senescent type 1 cells and innate immunity (Chapter 7).

The striking ultrastructural feature of the type 2 cell is the lamellar body. These are composed of stacked layers of membranes which contain the extremely important phospholipid–protein aggregate, surfactant. The composition of surfactant is complex and consists of phospholipids, neutral lipids and proteins (see Table 4.3). The latter includes surfactant proteins (SP) A, B, C and D. The most important constituent of surfactant that is responsible for the majority of its surface tension lowering effect is dipalmitoylphosphatidylcholine (DPCC). The presence of surfactant in the alveolar regions increases lung compliance markedly and thus reduces the work of breathing. It does this by reducing the surface tension of the alveoli at the air–liquid interface, preventing their collapse at low lung volumes and facilitating the ease with which lung inflation occurs (see Figure 4.8). Thus, alveolar stability is maintained across a wide range of lung volumes, allowing a more even matching of ventilation and perfusion. Surfactant also helps keep the alveolar space from flooding by maintaining a low surface tension in the interior of the alveolus, preventing interstitial fluid from being drawn inside. Global surfactant deficiency in premature human infants is the primary cause of respiratory distress syndrome and is now treated successfully with synthetic surfactant preparations [43]. Trials of surfactant preparations in adults in acquired respiratory distress syndrome have been less successful.

Surfactant production and catabolism is tightly regulated. Surfactant is produced in the rough endoplasmic reticulum and stored in the lamellar bodies of the type 2 cell (see Figure 4.9). The contents of the lamellar bodies including surfactant are extruded onto the air–liquid interface by a process of exocytosis. An intracellular influx of calcium triggers exocytosis in response to stretch of the alveolar region caused by hyperventilation or deep breathing. On the alveolar surface, the extracellular form of surfactant, tubular myelin, is formed. This step requires the presence of SP-A, SP-B and calcium. Tubular myelin exists as a bilayer at the air–liquid interface in the alveolus and serves as

TABLE 4.3: Composition of pulmonary surfactant

Phospholipids: 85%*	% of Phospholipids
Phosphatidylcholine	76.3
Dipalmitoylphosphatidylcholine	47.0
Unsaturated phosphatidylcholine	29.3
Phosphatidylglycerol	11.6
Phosphatidylinositol	3.9
Phosphatidylethanolamine	3.3
Sphingomyelin	1.5
Other	3.4
Neutral Lipids: 5%†	
Cholesterol, free fatty acids	
Proteins: 10%‡	
SP-A	
SP-B	
SP-C	
SP-D	
Other	

*The phospholipid composition is constant in most mammalian species. Disaturated phosphatidylcholine represents about two thirds of the total phosphatidylcholine. Dipalmitoylphosphatidylcholine makes up the majority species of the disaturated phosphatidylcholine fraction and is the critical molecule for providing the low surface tension.
†There is about 5% neutral lipid, of which the majority is cholesterol and free fatty acids. There is relatively little triglyceride and cholesterol ester.
‡The composition of the surfactant proteins is not known precisely, but on a mass basis, there appears to be more SP-A than SP-D and more SP-A than SP-B and SP-C.
(Reproduced with permission from Murray and Nadel's Textbook of Respiratory Medicine, 5th edition, 2010, p. 202, Philadelphia, Saunders Elsevier.)

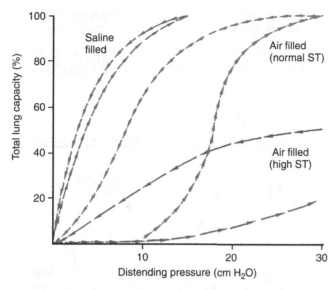

FIGURE 4.8: Pressure–volume curves. Pressure–volume curves of lungs filled with saline (blue) and with air with normal (green) and high (red) surface tension (ST). Arrows indicate inflation and deflation. When the lung is filled with saline, the effects of surface forces at the air–liquid interfaces are eliminated. The differences between the curve of the lung with normal surface tension (green) and that of the lung with high surface tension (red) are due to the reduction of surface forces by surfactant. (Modified and reproduced with permission from George RB et al. Chest Medicine: Essential of Pulmonary and Critical Care Medicine, 5th edition, 2005, p. 28, Philadelphia, Lippincott Williams and Wilkins.)

an extracellular reservoir of surfactant. It eventually adsorbs to the surface monolayer in the alveolus where it exerts the surface lowering tension effect that is vital to proper inflation/deflation cycles of the alveolar region. During deflation of the alveolus, the surface-active monolayer is compressed resulting in a rise in the liquid film pressure. Small aggregates of surfactant are squeezed out of the filmy monolayer to be recycled by nearby type 2 cells and resident alveolar macrophages. Macrophage catabolism of surfactant is regulated by granulocyte–macrophage colony stimulating factor (GM-CSF). The acquired form of a rare human lung disease called pulmonary alveolar proteinosis appears to be caused by a circulating autoantibody to GM-CSF [44]. In this disease, alveolar lung macrophages are unable to catabolize surfactant and related phospholipids. This results in a slow buildup of these materials inside the alveolus which then interferes with critical gas exchange functions causing shortness of breath and eventually respiratory failure. SP-B is also critical for proper surfactant function. Mouse models with homozygous null alleles for SP-B die of respiratory failure shortly after birth as do human infants homozygous for mutations in the SP-B gene [45, 46]. The

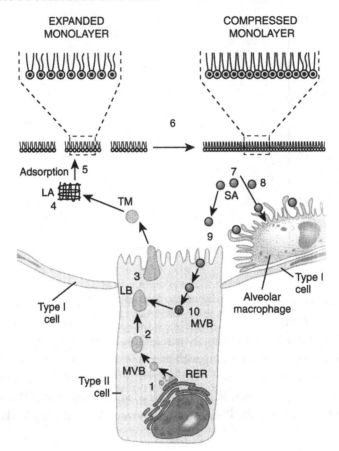

FIGURE 4.9: Surfactant synthesis. Surfactant phospholipids are synthesized in the rough endoplasmic reticulum (RER) and processed in the Golgi network of the alveolar type 2 cell (step 1). They are transported to the multivesicular bodies (MVB) (step 2), where lamellar bodies (LB) are formed. The lamellar bodies are secreted by exocytosis (step 3) onto the alveolar surface where they unfold to form tubular myelin (TM) (step 4) and other large aggregates (LA). Tubular myelin and the large aggregates adsorb into the expanded surface monolayer of the alveolus (step 5). This is a critical step for producing a low surface tension in the alveoli of the lung. During normal respiratory breathing, the expanded surface monolayer is compressed during exhalation causing a rise in film pressure which in turn creates a compressed monolayer of almost pure dipalmitoylphosphatidylcholine (step 6). Excluded material from the compressed monolayer (step 7) forms small aggregates (SA) which are either ingested by macrophages (step 8) or endocytosed and reprocessed by type 2 alveolar cells (steps 9 and 10). (Reproduced with permission from Murray and Nadel's Textbook of Respiratory Medicine, 5th edition, 2010, p. 208, Philadelphia, Saunders Elsevier.)

role of SP-C is incompletely understood. While it is found in tubular myelin, it does not appear to be essential for its formation. It likely plays a role in stabilizing the surface film of surfactant in the alveolus. In mouse models, the lack of SP-C is associated with chronic pneumonitis and alveolar enlargement at about 6 months of age [47]. Mutations of SP-C gene in humans have been associated with interstitial lung disease [48]. SP-D is not essential for surfactant function and appears to be associated with innate immunity in the lung [49]. SP-D as well as SP-A belong to a family of proteins known as collectins. Both SP-A and SP-D function as opsonins, binding to the surface of microorganisms and facilitating their clearance by phagocytes [50]. SP-D may regulate macrophage function in the alveolar region as well as assist in the clearance of apoptotic cells.

Type II cells synthesize many other proteins (see Table 4.2). Some are important growth factors such as VEGF (vascular endothelial growth factor) and cytokines such as IL-1 and IL-8. Receptors for growth factors are important components of type II cell structure, facilitating cell signaling necessary for epithelial cell proliferation and differentiation (see below). Essential matrix proteins such as laminin, fibronectin and type IV collagen are also produced. The presence of Na^+ and K^+ channels in type II cells as well as aquaporins such as AQP-3 and AQP-1, CFTR channels, and beta-adrenergic receptors emphasize the key role type II cells play in fluid and solute flux across the alveolar epithelium (see Chapter 6).

Type II cells have long been shown to play a major role in the restoration of normal alveolar architecture following lung injury [41]. Type II cells, when stimulated by certain growth factors, may divide to produce additional type II cells or transform into type I cells [51, 52]. Fibroblast growth factor receptor (FGF-R) and the receptor for hepatocyte growth factor c-Met enable these cells to respond to molecular signals and initiate proliferation of type II cells or transformation of type II cells into type I cells. The latter transformation appears to be reversible and both type II and type I cells may undergo apoptosis [53]. In addition, some *in vivo* human data exist, indicating that type II cells may also be able to differentiate into interstitial fibroblasts which may then play a role in lung repair and even pulmonary fibrosis [54]. Type II cell repair of the alveolar epithelium may be augmented by the recent discovery of true human lung stem cells present in the terminal airways and alveoli [55].

4.3.3 Alveolar Macrophages

The alveolar macrophage is the resident phagocytic cell of the alveolus and plays a major role in host defense (see Chapter 7.1.2). It can ingest both bacterial and particulate matter and secretes a number of cytokines that may recruit other cells to areas of inflammation. As discussed above, it is an essential cell in maintaining alveolar homeostasis through activities such as surfactant catabolism.

4.3.4 Interstitial Fibroblasts

The most abundant cells found in the alveolar region of the human lung are interstitial fibroblasts [56]. Although not found in alveolar walls, they are important cells because they are largely responsible for the secretion of the extracellular matrix that supports the alveolar units. When transformed into myofibroblasts under the stimulation of TGF-β, they likely play an important role in the development pulmonary fibrosis [57]. There is evidence that interstitial fibroblasts may arise from several different sources. Earlier studies indicated there was a resident population present from birth [58]. Later studies have suggested that bone marrow-derived fibrocytes may also be a contributing source [59, 60]. Finally, lung epithelial cells (see above) under stimulation by TGF-β may transform into fibroblasts [54].

4.3.5 Alveolar Capillary Endothelium

Alveolar capillary endothelial cells constitute the second most common cell type in the alveolar region of the lung (see Table 4.1). Unlike the glomerular endothelium of the kidney, the alveolar capillary endothelium is non-fenestrated and presents a tighter barrier against fluid and solute movement across the alveolar capillary membrane (see Chapter 6). It is bathed by the entire cardiac output of the right heart, giving it immense strategic importance. In recent years, important metabolic functions have been recognized and elucidated. The capillary endothelium of the lung, much like the Type 1 alveolar epithelium, is very elongated, thin and squamoid in appearance with a minimum of organelles. Adjoining cells are attached by specialized junctions called "tight" junctions (regulated by the protein occludin) and "adherens" junctions (Ca^{++}-dependent junctions regulated by the cadherin family of membrane proteins) [61]. A very prominent feature (also seen in the alveolar epithelium) is the presence of numerous caveolae (see Figure 4.7). These structures likely play a role in transport of macromolecules such as albumin as well as increasing the surface area of the endothelium, thus providing additional area for important surface active enzymes such as angiotensin-converting enzyme.

The pulmonary capillary endothelium resembles the pulmonary vascular endothelium of the larger vessels (>25 μm) by the presence of typical markers such as LDL (low density lipoprotein) uptake, eNOS (endothelial nitric oxide synthase), factor VIII von Willebrand antigen, VE-cadherin (vascular endothelial cadherin) and PECAM-1 (platelet-endothelial cell adhesion molecule) [62]. The pulmonary capillary endothelium differs from the larger pulmonary vascular endothelium in several important respects. The capillary endothelium presents a tighter barrier against fluid and solute exchange compared to the relatively "leaky" endothelium of larger vessels [63]. Proliferation of capillary endothelial cells is faster than their larger vessel counterparts, likely a useful feature given the delicate and strategic importance of their location [64]. Pulmonary capillary endothelium does not flow align as large vessel endothelium does and capillary endothelium exhibits distinctive

calcium and oxidant-mediated signal transduction mechanisms compared to its larger vessel coun-
terpart [65, 66].

4.4 PULMONARY CIRCULATION AND DISTRIBUTION OF BLOOD FLOW

The exterior walls of the alveoli are primarily composed of small capillaries (thus forming the alveolar–capillary membrane as discussed later) derived from branches of the pulmonary artery (see Figure 4.10). The alveolar walls are delicate structures, ideal for gas exchange, but may be susceptible to
injury from various insults which result in overdistention of the alveolus, such as barotrauma. This
rich meshwork of capillaries drains into nearby pulmonary venules, located at the periphery of the
terminal respiratory unit. Venules join up to form pulmonary veins which in turn eventually enter
the left atrium of the heart.

The total volume of blood contained in the pulmonary circulation is about 450 mL, representing approximately 12% of the total blood volume in an average sized human body [67]. In

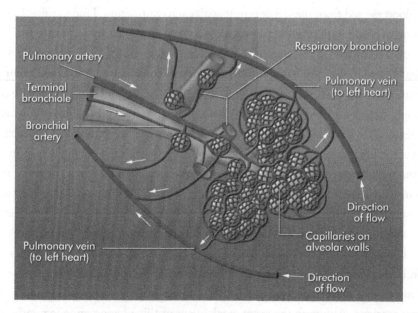

FIGURE 4.10: Pulmonary circulation to the terminal respiratory unit. The pulmonary artery runs parallel to the airways and breaks up into a capillary bed over the alveolar region where gas exchange occurs.
The bronchial artery (derived from the aorta) supplies blood to the airways to the level of the respiratory
bronchiole. It also supplies blood to the visceral pleura. The pulmonary veins lie within the substance of
the lung and drain the alveolar capillary network, the airways, and the visceral pleura.

addition to its main role in gas exchange, the pulmonary circulation has several other important roles in maintaining physiologic homeostasis in the normal human lung. The pulmonary circulation serves as an important reservoir for blood to the left ventricle and can sustain left ventricular stroke volume for several beats if systemic venous return is transiently less than left ventricular output. Similar to the situation in the small airways, the total cross sectional area of the pulmonary capillaries is many times larger than the pulmonary arteries and veins. Since virtually all of the mixed venous blood returning to the heart passes through the pulmonary capillaries, these tiny vessels may serve as a filter to keep various particles such as small blood clots, tumor cells, fat cells, gas bubbles, collections of platelets and leukocytes and other debris from entering the systemic circulation. The pulmonary circulation also has important metabolic activity in the synthesis, conversion or clearance of a number of vasoactive substances. The peptide angiotensin I is converted to the potent vasoconstrictor angiotensin II by the pulmonary endothelium [68]. Other vasoactive substances such as bradykinin, serotonin, leukotrienes, endothelin and prostaglandins E_1, E_2 and $F_{2\alpha}$ are almost completely removed. Nitric oxide, a powerful vasodilator, is produced in the pulmonary endothelium by nitric oxide synthase [69].

Normally about 70 mL of blood is contained within the pulmonary capillaries, about the same as the blood ejected from the left ventricle with each contraction. Due to the effects of gravity, the distribution of blood in the capillary network is uneven with more blood contained in the capillaries of the bases compared to the apex of the lung (see Figure 4.11). This has the effect of distending capillaries located in the lung bases and therefore decreasing resistance to flow (compared to the apical portions of the less perfused capillary bed). Furthermore, under conditions where cardiac output is increased, such as vigorous exercise, less distended capillaries are recruited, allowing an almost fourfold increase in capillary blood volume while maintaining a low pulmonary vascular resistance. These two basic mechanisms of capillary distension and recruitment are largely responsible for the decrease in pulmonary vascular resistance as pulmonary arterial pressure rises (such as exercise and other states in which cardiac output is increased) and are crucial in maintaining the relatively low vascular pressures seen in the pulmonary circulation compared to the systemic circulation.

Other mechanisms exist to match ventilation and perfusion besides capillary distension and recruitment. Hypoxic pulmonary vasoconstriction occurs in the pulmonary arterioles of the precapillary bed in response to alveolar hypoxia. Smooth muscle in the walls of these vessels constricts when a fall in ambient alveolar oxygen is detected and blood flow is rerouted to other alveoli with higher oxygen tensions. Although the exact mechanism of this phenomenon is poorly understood, it is clear that a fall in the alveolar oxygen tension and not the pulmonary artery oxygen tension is the key initiating event. This has been demonstrated by perfusing lungs with normoxic pulmonary arterial blood and hypoxic alveolar gas while observing vasoconstriction.

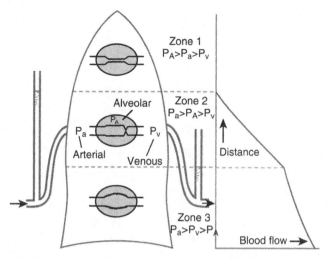

FIGURE 4.11: West's zones of pulmonary blood flow. The three-zone model used to describe pulmonary blood flow through the alveolar regions of the lung. Although oversimplified, this model is useful in conceptualizing the complex physiology that determines perfusion of the alveoli. In short, pulmonary capillary blood flow increases as distance from the lung apex increases, mostly due to the effects of gravity. P_A, pulmonary alveolar pressure; P_a, pulmonary arterial pressure; P_v, pulmonary venous pressure. In the apex of the lung (Zone 1) where P_A is greater than P_a and P_v, very little blood flow occurs through the alveolar capillaries. At the base of the lung by contrast (Zone 3), P_a is greater than both P_v and P_A, and thus capillary blood flow is maximal. Zone 2 is the intermediate zone where blood flow is determined by the difference between P_a and P_A (frequently referred to as a "Starling resistor" phenomenon in which a collapsible tube is surrounded by a pressure chamber). (Reproduced with permission from Murray and Nadel's Textbook of Respiratory Medicine, 5th edition, 2010, p. 63, Philadelphia, Saunders Elsevier.)

Capillary blood flow around the alveoli is normally pulsatile with a velocity of around 1000 μm/second. Red blood cells in the capillaries transit the alveolar region (usually crossing several alveoli) to reach the venules in about 0.75 seconds. Under normal circumstances, diffusion of oxygen and carbon dioxide across the capillary endothelium and alveolar epithelium is very rapid and is not limited by this transit time. However, under conditions of extreme exercise, low ambient oxygen concentration or lung disease diffusion limitations may exist.

· · · ·

CHAPTER 5

The Alveolar–Capillary Membrane and Gas Exchange

The alveolar–capillary membrane is the key region of the gas exchange area in the TRU. Here, diffusion of oxygen from the alveolus to the red blood cell of the alveolar capillary takes place. The reverse situation occurs for carbon dioxide. Diffusion may be defined simply as the tendency of molecules to move from a region of higher concentration to a region of lower concentration. Fick's law states that the diffusion of a gas is directly proportional to the surface area for gas exchange and inversely proportional to the length of the diffusion pathway. Thus, the literal "sheet of blood" (almost 80 m^2) that surrounds the alveoli along with the short distance from the inside of the alveolus to the inside of the capillary (about 0.5 µm), are optimal for efficient diffusion of oxygen and carbon dioxide. The alveolar–capillary membrane may be conceptualized as a series of barriers through which oxygen and carbon dioxide must diffuse (see Figure 5.1). These include: 1) fluid lining the inside surface of the alveolus, 2) the alveolar epithelium, 3) the alveolar interstitial space 4) the capillary endothelium 5) the plasma layer in the capillary and 6) the red blood cell membrane. Both carbon dioxide and oxygen diffuse very rapidly in a non-liquid medium such as air compared to water. Fresh ambient atmospheric gas contained in the conducting airways moves by bulk flow to the region of the terminal bronchiole where gas velocity rapidly slows to the point that diffusion becomes the predominant means of airflow. Freshly inspired oxygen rich air rapidly diffuses into the alveoli and thence across the alveolar capillary membrane into the red blood cell. Only a small amount of oxygen remains dissolved in the plasma of the alveolar capillary. The vast majority diffuses across the red blood cell membrane to combine with the four hemoglobin binding sites. These become fully saturated with oxygen in 0.25 seconds or less, providing a substantial diffusion reserve capacity since the transit time of the red cell across the alveolar regions is about 0.75 seconds. The diffusion of oxygen from the alveolus to the red blood cell is also facilitated by the large difference in the partial pressure of oxygen between the alveolus and the pulmonary capillary blood. This difference persists even after all red cells have been fully loaded with oxygen in their transit across the alveolar region of the TRU.

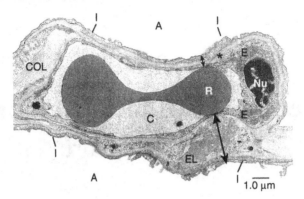

FIGURE 5.1: The alveolar–capillary membrane. Cross-section of the alveolar–capillary membrane showing the path of diffusion for oxygen and carbon dioxide. A, alveolus; R, red blood cell; C, plasma in the capillary; E, endothelial cell; Nu, nucleus of endothelial cell; I, type 1 alveolar epithelial cell.*, interstitium formed by fused basal laminae of endothelial and epithelial cells. COL, collagen; EL, elastin. The short double arrow demonstrates the "thin side" of the alveolar–capillary membrane through which gas diffusion most easily occurs. The long double arrow illustrates the "thick side" of the membrane through which most solute and fluid exchange occurs (see Chapter 6). (Human lung surgical specimen, transmission electron microscopy, Reproduced with permission from Murray and Nadel's Textbook of Respiratory Medicine, 5th edition, 2010, p. 14, Philadelphia, Saunders Elsevier.)

Carbon dioxide is about 20 times more soluble in water and therefore more diffusible than oxygen. It is carried through the alveolar capillaries in 3 forms: approximately 5%, is physically dissolved in the plasma. The remainder exists within the red blood cell in two forms: 5% is bound to hemoglobin as carbamino compounds and 90% exists as bicarbonate (HCO_3^-). As blood flows through the pulmonary capillaries surrounding the alveoli, oxygen quickly diffuses from the alveoli into the capillaries. Almost all of the oxygen is bound to the hemoglobin molecule which displaces the majority of the carbon dioxide contained in the red cell. The carbon dioxide then diffuses across the alveolar capillary membrane and into the alveolus and then along the airways to be expelled into the atmosphere. The loss of carbon dioxide from the red blood cell results in a decrease in hydrogen ion (H^+) concentration in the cell (pH increases). Thus, oxygenated blood has a reduced capacity for carrying carbon dioxide (at the alveolar level). The effect of changes in blood CO_2 concentrations and intracellular pH on hemoglobin O_2 affinity is called the Bohr effect (see Figure 5.2). Conversely, deoxygenated blood (at the tissue level) has an increased capacity for loading carbon dioxide. This important physiological phenomenon is called the Haldane effect.

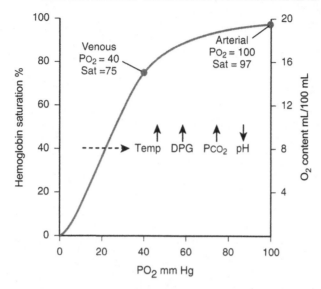

FIGURE 5.2: Oxygen–hemoglobin dissociation curve. Oxygen–hemoglobin dissociation curve demonstrating values typical for mixed venous blood and arterial blood. Note that the curve is shifted to the right by increases in temperature, PCO_2, 2,3-diphosphogycerate (DPG), and H^+ concentration. Sat is hemoglobin saturation (expressed in %). (Reproduced with permission from Murray and Nadel's Textbook of Respiratory Medicine, 5th edition, 2010, p. 85, Philadelphia, Saunders Elsevier.)

CHAPTER 6

Fluid and Solute Exchange

The alveolar epithelium and the alveolar capillary endothelium control the flux of water and solute material across the alveolar/capillary membrane. This is where the majority of fluid and solute exchange occurs in the human lung. The alveolar epithelium constitutes a much tighter barrier and is less permeable to water and solutes than the capillary endothelium [70]. Colloid osmotic pressure, created by the plasma proteins of the surrounding capillaries, assists in keeping the alveoli relatively fluid free (except for the thin lining of surfactant). Colloid osmotic pressure is usually greater than capillary hydrostatic pressure and tends to pull excess fluid from the interior of the alveoli. Similar to most organs, there is a net outward flow of fluid from the capillary endothelium. This fluid is drained by nearby lymphatics in the surrounding interstitium and moves toward the hilum of the lung by the slightly negative interstitial pressure gradient. Fluid flux from the alveolar capillary vessels into the alveolar interstitial compartment is classically described by the Starling equation for fluid filtration across a semipermeable membrane:

$$\dot{Q} = K[(P_{mv} - P_{pmv}) - (\pi_{mv} - \pi_{pmv})]$$

$\dot{Q}$ is the net transvascular fluid flow, K describes the permeability of the membrane, P_{mv} is the hydrostatic pressure inside the capillaries, P_{pmv} is the hydrostatic pressure of the surrounding interstitial space, π_{mv} is the plasma protein osmotic pressure inside the capillaries, and π_{pmv} is the plasma protein osmotic pressure of the surrounding interstitial space. P_{pmv} is very low, close to alveolar pressure (zero or atmospheric pressure), thus the main hydrostatic force for fluid filtration is P_{mv} and this favors net fluid filtration into the interstitial space. P_{mv} varies depending on the height of the lung (Zones 1, 2 or 3 conditions as previously described). The net outflow of fluid from the capillaries created by hydrostatic pressure is partially counterbalanced by the protein osmotic pressure, which is greater inside the capillaries than in the surrounding interstitial space ($\pi_{mv} > \pi_{pmv}$). In the normal human lung, approximately 10–20 mL/hour of fluid is filtered and removed by the lymphatics and then returned to the systemic circulation [71].

Due to the extremely tight alveolar epithelial cell junctions, very little if any fluid normally enters the alveolus from the interstitial compartment [72, 73]. Oxygen and carbon dioxide, which

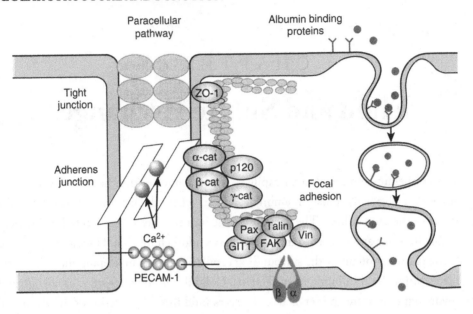

FIGURE 6.1: Fluid and solute transport. Schematic depiction of paracellular and transcellular transport mechanisms for fluid and solute exchange across the microvascular endothelium at the alveolar level. Albumin transport may occur via transcellular mechanisms in which albumin binding proteins on the luminal surface bind circulating albumin. Vesicles are formed and albumin is transported through the endothelial cell. Vesicles fuse to the abluminal surface and albumin is then extruded on the abluminal side. The paracellular pathway is controlled by a series of junctions including PECAM-1 (platelet-endothelial cell adhesion molecule-1) junctions, tight junctions, and adherens junctions. The tight junctions consist of proteins linked to the cell's actin cytoskeleton by the zona occludens family (ZO-1). The adherens junctions are controlled by Ca^+ dependent cadherins proteins which in turn are linked to the cell's α-, β-, and γ-catenin (cat) complex. Focal adhesion plaques consisting of α- and β-transmembrane proteins are linked to the actin cytoskeleton by a complex of proteins including paxillin (Pax), vinculin (Vin), talin, and focal adhesion kinase (FAK). These adhesion plaques are important in cell-matrix tethering. (Reproduced with permission from Murray and Nadel's Textbook of Respiratory Medicine, 5th edition, 2010, p. 122, Philadelphia, Saunders Elsevier.)

are lipid soluble and of low molecular weight, are freely permeable across the alveolar epithelium as well as the endothelium. Diffusion of these gases takes place primarily across the thin portion of the alveolar/capillary membrane, an area where endothelial cells and alveolar epithelial cells are practically adherent to each other. In contrast, fluid and solute exchange across the endothelium preferentially occurs in the thickest portion of the alveolar–capillary septum, an area that is more compliant (see Figure 5.1). The extracellular matrix proteins of this region consist of type I and type IV collagen, laminin, vitronectin, fibronectin, and proteoglycans [74]. This matrix also serves as a barrier to solute transport and can sieve molecules of different sizes.

Paracellular and transcellular pathways exist for the movement of water and solutes across the microvascular endothelium (see Figure 6.1). Water moves freely across the endothelium via aquaporin water channels [75]. Transendothelial channels are also available for water transport as well as low-molecular weight lipid-soluble substances. Vesicles formed from caveolae are available for the transport of macromolecules such as albumin. Diffusion of larger molecules such as proteins is restricted to intercellular junctions. Passage of molecules through intercellular junctions is regulated by a variety of junctional proteins forming adherens junctions and tight junctions as well as the electrostatic charge of the molecules.

•   •   •   •

CHAPTER 7

Alveolar Defense and Clearance Mechanisms

There are two basic types of immunity that our lungs use in order to provide this defense, innate and adaptive. The innate immune system represents a non-specific form of defense whereas adaptive immunity is an antigen-specific defense. The innate system is mediated by structural and mechanical mechanisms, antimicrobial molecules and phagocytic defenses whereas the adaptive system uses cell-mediated and antibody-mediated mechanisms directed towards very specific invaders [76]. Our lung defenses go one step further and even provide a means of communication between these two types of immunity.

7.1   INNATE IMMUNITY

There are various soluble proteins and structural components that are the workhorses of the innate immune system. These proteins and structural components have specific purposes that aim to either identify invaders or recruit/activate other immune mediators that are the effectors of immunity. In order to more completely discuss the defense mechanisms of the lungs, we need to first define some of the components (see Table 7.1). The primary functions of the innate immune system are to activate cytokines and the complement cascade to direct leukocytes and antibodies to deal with invaders, if deemed appropriate. Both the upper airways and the alveolar spaces have the means to provide immunity. The upper airways have mechanical and physical barriers as well as cellular components. The mechanical and physical barriers include the cough reflex, mucous and mucociliary transport. The cellular defenses include the airway epithelium itself, specific types of immunoglobulins located in the mucosal surface and several types of lymphocytes all involved in the killing or inactivation of invaders.

Particulate matter that is inhaled may be deposited anywhere from the upper airways to the alveoli in the TRU. Generally, particles larger than 10 μm impact in the upper airway while particles between 2 and 10 μm impact in the lower portions of the bronchial tree. Particles between 0.5 and 3 μm may enter the gas exchanging areas in the TRU [77, 78]. Once in the alveolar regions of the lung, there are basically three components of the clearance system that work together to maintain

	TABLE 7.1: Innate immune system definitions.	
MEDIATOR	**DEFINITION**	**FUNCTION**
Cytokines and chemokines	Small cell-signaling protein molecules that are secreted by numerous cell types.	Recruit immune cells to sites of infection via production of chemical factors (i.e., interleukins)
Complement cascade	Small proteins that can become enzymes to cleave other proteins involved in cytokine production.	To "complement" the ability of antibodies and phagocytic cells to clear pathogens by formation of a membrane attack complex to poke holes in target cells, leading to cell lysis and death.
Leukocyte (i.e., white blood cells)	Effector cells of the body's immune system.	Seven cell types each with specific functions (neutrophil, lymphocyte, eosinophil, basophil, monocyte, macrophage and dendritic cell)
Antibody (i.e., immunoglobulins)	Large, Y-shaped proteins produced by specific lymphocytes to identify and neutralize foreign substances.	Tags a substance for attack by the immune system or directly neutralize it.
Antigen	Substance-specific proteins	Evokes the production of one or more antibodies.

(Adapted from Innate Immune System. 2012 [April 20, 2012]; Available from: http://en.wikipedia.org/w/index.php?title=Innate _immune_system&oldid=488271374.)

alveolar integrity. Alveolar macrophages can ingest particulate matter and then migrate to the respiratory bronchioles where ciliated epithelial cells clear them from the airways via the mucociliary escalator. The mucociliary elevator is composed of two layers of mucous (inner and outer layers) that sit atop the ciliated airway cells. This mucous is secreted largely by airway goblet cells with additions from the surfactant secreted by the type II alveolar epithelium. The inner layer of mucous, the sol layer, is thin enough to permit sustained, rhythmic beating by the cilia. The top layer, the gel layer, is stiffer and entraps particles and macrophages and gradually moves them up the airways where they are eventually coughed out or swallowed [79]. Macrophages may also penetrate into the interstitial areas near the alveoli where they enter the lymphatics of the deep plexus. These lymphatic channels do not directly surround the alveoli, rather they are located in the interstitial spaces near the terminal bronchiole and the associated pulmonary arteries and veins [80]. Macrophages containing particulate matter are moved through these channels to regional lymph nodes surrounding the major bronchi and trachea.

7.1.1   Alveolar Epithelium

The alveolar epithelium, composed of type I and type II pneumocytes (also called alveolar epithelial cells), represents the first line of defense against invading pathogens and toxins. These cells form a physical barrier to prevent these invaders from penetrating into the interstitium of our lungs. As previously discussed, type I pneumocytes comprise about 95% of the alveolar surface and are mainly involved in gas exchange. The role of the type I pneumocyte in immunity seems to be limited. Although the type II pneumocyte covers only 5% of the alveolar surface area, it has a very large role in pulmonary immunity [81]. The type II pneumocytes are responsible for surfactant production, help in reforming the alveolar epithelium after damage, transport sodium and fluid into the interstitium and have an important role in the innate immunity of the lungs. The role of the type II pneumocyte in innate immunity will be the focus of this chapter.

One of the major functions of the type II pneumocytes is production of various soluble proteins that line the alveolar walls, including surfactant (see Chapter 4.3.2). In addition to its previously discussed role in improving lung mechanics, surfactant has antimicrobial and anti-inflammatory properties as well and may even have a role in suppression of an inappropriate inflammatory response. SP-A and SP-D, together with mannose-binding protein, conglutinin and the complement cascade member C1q, make up a group of proteins called collectins that appear to have their primary role in innate immunity [81, 82, 83]. These collectins are primarily involved in enhancing microbial phagocytosis by macrophages by acting either as an opsonin, a molecule used to tag an antigen, or directly stimulating the macrophage [84]. SP-A has also been shown to help control or decrease the inflammatory response through interactions with a specific receptor present on the surface of alveolar macrophages [83]. The main difference between SP-A and SP-D is the types of antigens that are targeted and the mechanisms they use to enhance the immune function. Together, they target a wide range of respiratory pathogens including common bacteria, such as *Streptococcus pneumoniae* and *Staphylococcus aureus*, less common bacteria like various *Mycobacterium* species and even some viruses including herpes simplex I and respiratory syncytial virus [83].

There are several other proteins produced by type II pneumocytes that are involved in innate immunity. Each of these proteins has specific functions directed against various types of microorganisms. Lipopolysaccharide binding protein (LBP), produced in the liver, has recently been found to be produced by pulmonary artery smooth muscle cells as well as type II pneumocytes [85, 86]. LBP works by stripping some of the lipopolysaccharide (LPS) from the outer membrane of gram-negative bacteria for recognition by specific leukocytes [82]. There is an entire family of protein receptors, called the Toll-like receptors (TLRs), located on the surface of alveolar macrophages as well as epithelial cells that mediate the interaction between these LBP:LPS complexes and the specific receptors on the leukocytes. It is not just the LBP:LPS complexes that interact with the TLRs however. Several different proteins involved in innate immunity interact with these receptors to mediate the killing of various bacteria, viruses and fungi.

Other proteins involved in innate immunity have either direct antimicrobial activity or facilitate phagocytosis. These other proteins include lysozyme, which kills bacteria by forming pores in the their membranes; lactoferrin, which excludes iron from bacterial metabolism; immunoglobulins A and G (IgA and IgG); and defensins, which are antimicrobial peptides released from leukocytes and epithelial cells that, like lysozyme, form pores in microbial membranes [82]. Much of the activity of the alveolar epithelial cells is directed toward producing proteins that tag and signal activation of other phagocytic cells such as the alveolar macrophage (discussed in the next chapter). Macrophages, in addition to ingesting microbial proteins, also signal other cells to initiate an inflammatory cascade in a specific area of the lung via the production of various cytokines and chemokines. Alveolar epithelial cells can produce these cytokines and chemokines as well, thereby aiding in the innate inflammatory response in the distal airways [87].

7.1.2 Monocytes and Alveolar Macrophages

There are many specific cells types located in the alveolar space that help provide our innate immunity; none however are more abundant than the offspring of the monocyte, the alveolar macrophage. Monocytes, produced in the bone marrow from the granulocyte/monocyte progenitor cell, are responsible for a large part of this defense mechanism. Not only are they the progenitor of macrophages, but they also differentiate into dendritic cells (discussed in Chapter 7.2.2).

Alveolar macrophages reside in the epithelial lining fluid and have several functions. They catabolize pulmonary surfactant and are avidly phagocytic. These cells are able to ingest all types of particulates (microbial and non-microbial, harmful and harmless) without necessarily provoking a full inflammatory response. In fact, alveolar macrophages are even able to suppress an inappropriate inflammatory or immune response. Once the alveolar macrophage ingests a particle, infectious or non-infectious, it becomes enclosed within a phagosome that eventually fuses with a lysosome. Lysosomes are basically the digestive system of the cell. They contain specific enzymes that breakdown material and cellular debris. Most material encountered by the alveolar macrophage can be broken down, but some bacterial components and inorganic materials are not. These substances stay sequestered in the lysosome where they remain for the life of the macrophage [88]. These particle-laden macrophages have several potential fates. Some of them will be cleared via the mucociliary escalator, others drain to regional lymph nodes, while others remain in the lung for several months only to release their particles when the macrophage dies [89, 90]. The particles are then taken up by other macrophages.

Materials that are inhaled must be tagged in order for the alveolar macrophage to recognize them. This is done via several of the previously mentioned proteins such as the TLRs, SP-A and SP-D, as well as various members of the compliment cascade all of which the alveolar macrophage has receptors for on its cell surface. Once the alveolar macrophage recognizes and ingests an inhaled particulate, it has several mechanisms at its disposal by which to deal with them. Some of these mechanisms involve directly killing the invading microbe via generation of reactive oxygen species (ROS) or nitric oxide (NO) while others augment the innate immune response via the production of cytokines and chemokines.

If the alveolar macrophage is not able to digest the foreign material itself, it produces chemicals called cytokines and chemokines in order to recruit other inflammatory cells or leukocytes. There are several types of leukocytes in the armamentarium of our immune system (see Table 7.1), and each have a specific function. The first cells recruited to an area of inflammation are neutrophils, followed by monocytes and lymphocytes [88]. Some of the products produced by the alveolar macrophage have functions that not only signal for other leukocytes, but also promote leukocytic activity or make it easier to find the areas in need of an immune response. These products are called lipid mediators and are produced by the alveolar macrophage itself in response to activation of one of the TLRs present on its cell surface. One of these mediators is platelet-activating factor (PAF), produced by the acetylation of lysophosphatidylcholine, whose actions lead to an increase in the adhesive properties of neutrophils to vascular endothelial cells as well as bronchoconstrictive and vasodilative activities [91, 88]. Other products of the alveolar macrophage, produced by the oxidation of arachidonic acid by the cyclooxygenases 1 and 2 (COX-1 and COX-2), include the prostaglandins, more specifically prostaglandins E_2, D_2 and I_2 (PGE_2, PGD_2 and PGI_2) and thromboxane A_2 [92]. These molecules have a role in the regulation of vascular tone and in the inhibition of macrophage function. One of the more important arachidonic acid products, leukotriene B_4 (LTB_4), produced by 5'-lipoxygenase, is the primary factor responsible for recruitment of neutrophils early in the inflammatory response [93].

These lipid mediators are vital to the start of an inflammatory response, but it is the formation of cytokines and chemokines that keep it going. There are several cytokines that either promote or inhibit the inflammatory response and although many will be listed here, it is by no means exhaustive. The most abundant cytokines produced by alveolar macrophages are those with antimicrobial and antiviral activity. These include the aforementioned ROS and NO, but also include tumor necrosis factor-alpha (TNF-α) and interferon-alpha and beta (IFN-α, IFN-β) [94]. The mechanisms by which these factors produce microbial killing are complex, but the end result is to interfere with the replication of microbial cells, induce apoptosis or signal the production of other factors leading to a systemic inflammatory response. Other cytokines are more involved in the attraction of neutrophils. These include interleukin (IL)-8, macrophage inflammatory proteins 1 and 2 (MIP-1 and MIP-2), LTB_4 and platelet-derived growth factor (PDGF) [95, 96, 97]. Some of the interleukins (IL-1α, IL-1β, IL-6) and TGF-β are also involved in non-specific resistance against infective agents [94, 98]. There are still other cytokines that have fibrogenic activity via the stimulation of fibroblasts including PDGF, TGF-β, FGF, TNF and fibronectin [99, 100]. Alveolar architecture is built on a matrix of connective tissue that resides in the extracellular space and macrophages contribute to both the synthesis and degradation of it. Non-functional areas of collagen and elastin are broken down by matrix metalloproteinases (MMPs) while the production of FGF, PDGF and other cytokines promote the formation collagen and blood vessels while simultaneously inhibiting further expression of the MMPs.

Although cytokines and chemokines have been lumped together, they are not necessarily synonymous. Chemokines are chemicals that act as cytokines to guide the migration of other cells, hence chem-okines. There are four families of chemokines based on their small size and the presence of a cysteine residue in their molecules called C, CC, CXC and CXXXC [94,101]. It is the CXC and CC families that play an important role in inflammation in the lung. The CXC family is involved in neutrophil attraction via IL-8, growth regulated oncogene (GRO)-α, -β and -γ, epithelial cell-derived neutrophil-activating peptide (ENA)-78 and others. The CC family attracts mononuclear cells (monocytes and lymphocytes) via several factors including monocyte-chemotactic protein (MCP)-1 and RANTES (regulated upon activation, normal T-cell expressed and secreted).

The ability of the alveolar macrophage to suppress inflammation is a characteristic not present in macrophages found in other parts of the body [102]. As noted earlier, the interaction of SP-A with a specific receptor present on the surface of the alveolar macrophage is responsible for this function. That receptor is called the signal inhibitor regulatory protein (SIRP)-α, and its actions mediated the integrin $\alpha v \beta 6$, IL-10, PGE_2 and NO [83, 103, 104]. These last three factors also play a role in suppressing some of the activity of the adaptive immune system by down-regulating the antigen-presenting activity of dendritic cells [105, 106].

In summary, the alveolar macrophage plays a key role in our innate immune system, as well as our adaptive immune system (see Chapter 7.3). This cell is our first line of defense against invasion from various particles including bacteria, fungi, viruses and environmental particulates. It is through all of these mechanisms, lipid mediators, cytokines, intracellular components, etc., that the macrophage not only initiates an inflammatory response, but also maintains and moderates it to help keep us healthy.

7.1.3   Alveolar Leukocytes

There are several types of white blood cells (WBCs), or leukocytes, located in the alveolar space. The most prominent are the lymphocytes, but even these only comprise about 10% of the cells obtained on bronchoalveolar lavage (BAL) [107]. These are further divided into T-cells and B-cells and have their primary role in adaptive immunity. Other, less common leukocytes found in the alveolar space include eosinophils and basophils each representing <1% of cells recovered from BAL. The eosinophils are primarily involved in atopic reactions in response to T-lymphocytes and in the response to parasitic and fungal infections. The role for basophils in the alveolar space is less clear, but they may have a role in the late-phase bronchial responsiveness seen in asthma [108].

If the innate immune system represents a sword used to defend our lungs against invaders, then the alveolar epithelial cells and macrophages would be the blade. The tip however would be made from the neutrophils. These cells represent only about 2% of the cells present on BAL, but their numbers can increase substantially in response to invasion [107]. These cells, produced in the bone marrow, circulate throughout the body and migrate to sites of active inflammation in response to signals sent out by innate immune mechanisms as previously described.

Neutrophils gain access to sites of alveolar inflammation through the capillaries of the micro-circulation. Unlike other areas of the body where post capillary venules provide the conduit for neutrophil migration to sites of inflammation, neutrophils gain access to the alveolar regions of the lung only by squeezing through the tight spaces of the alveolar capillaries. This requires a shape change in the neutrophils, causing them to slow down in their transit and creating a reservoir of cells. When these neutrophils sense the presence of certain cytokines they stiffen, thereby promoting increased entrapment in the pulmonary capillaries [102, 109]. They then migrate between the endothelial cells before encountering the tight epithelial cell junctions. Despite these tight junctions, the neutrophil enters the alveolar space where they then become fully activated [110]. It is these tight junctions that prevent fluid from seeping into the interstitial spaces. There are some situations in which the neutrophils become activated prior to entering the alveolar spaces such as in severe sepsis. Premature neutrophil activation may damage both the endothelial and epithelial cells, decreasing the effectiveness of the intercellular junctions and resulting in pulmonary edema [111].

There are multiple mechanisms available to the neutrophil to kill invading microbes that are beyond the scope of this review. Some of these mechanisms include phagosome formation with the production of nicotinamide adenine dinucleotide phosphate (NADPH)-induced oxidative burst resulting in the formation ROS while others involve the production of defensins like human neutrophil proteins (HNP)-1, -2 and -3. Neutrophils can even project uncoiled nuclear DNA into the surrounding environment to form NETs (neutrophil extracellular traps) which contain cationic proteins including histones, defensins and cathelicidin [112, 113] (see Figure 7.1). Neutrophils also signal for the activation of adaptive immunity by recruiting monocytes, T-cells and dendritic cells

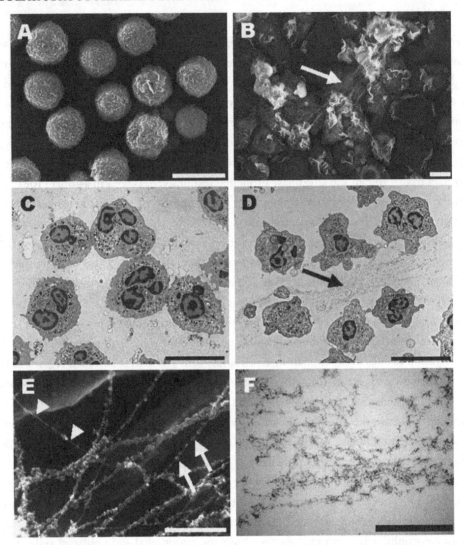

FIGURE 7.1: Neutrophil extracellular traps (NETs). Electron microscopic view of resting and activated neutrophils. Bars in A–D represent 10 μm and in E-F represent 500 μm. (Reproduced with permission from Brinkmann, V., et al., *Neutrophil Extracellular Traps Kill Bacteria*. Science, 2004. 303(5663): pp. 1532–1535.)

(A) Resting neutrophils are round and devoid of fibers.

(B) Upon stimulation, the cells flatten and make many membrane protrusions forming fibers (NETs), arrows in (B) and (D).

(C) Transmission electron microscopic analysis of naïve neutrophils

(D) Ultrathin section of stimulated neutrophils and several pseudopods and NETs (arrows)

(E) High-resolution scanning electron microscopic analysis of NETs that consist of smooth fibers (diameters of 15–17 nm, arrowheads) and globular domains (diameter around 25 nm, arrow).

(F) Ultrathin sections of NETs show that they are not membrane-bound.

[102]. Some of the products of neutrophils participate in the down-regulation of the inflammatory response as well. These factors include secretory leukocyte protease inhibitor (SLPI), which inhibits neutrophil elastase and lipoxins that inhibit neutrophil recruitment and ROS generation as well as enhance the uptake of apoptotic neutrophils by macrophages [114, 115].

7.2 BRIDGING THE GAP

Our innate immune system has developed in order to recognize pathogen-associated molecular proteins (PAMPs) including complex lipids, carbohydrates and certain types of DNA and RNA sequences present in and on microbial invaders. The activation of innate immunity can occur in as short a time period as three to four hours. The adaptive immune system responds to specific antigens. This system has the benefit of being able to remember specific antigens that have been presented to it previously and therefore mount a more rapid response on repeat exposure. Neither system can operate in isolation. An innate immune reaction occurs quickly, but the full response of the adaptive immune system may take up to 10 days [116]. Each is dependent on the other in order to maintain the integrity of our alveolar defenses.

7.2.1 Brief Overview of Immunology

This book is intended to be a relatively brief overview of the structure and function of the alveoli. This chapter specifically focuses on alveolar elements that participate in immunity. In order to further discuss adaptive immunity, some concepts of immunology need to be defined. This chapter is a concise outline of how our immune system carries out its tasks of cellular- and antibody-mediated immunity and the players involved in these processes.

Our immune system consists of several different types of cells, soluble proteins and organs in order to deal with foreign invaders. These components are present throughout our body, including the lungs and alveolar spaces. All of the cells that make up our immune system are derived from the subset of leukocytes called lymphocytes. These lymphocytes are further broken down into two main subsets called the T- and B-lymphocytes, so named because their development in the thymus or bone marrow, respectively. There is a third lymphocyte subset called natural killer (NK)-cells that does not express cell-surface markers categorized under the T- or B-lymphocyte subsets. It is the cell-surface markers that enable researchers to differentiate cell types based on their molecular composition called clusters of differentiation (CD).

The T-lymphocytes, or T-cells, are identified by the presence of a T-cell receptor for antigen (TCR) that is composed of either an αβ-chain or a γδ-chain. The exception to this is the NK-cell since they not only lack a T-cell CD marker but also lack a TCR. NK-cells respond to foreign invaders in a non-specific manner via Ig-receptors and production of certain cytokines [117]. If antigen is presented to a T-cell with a TCR the process of differentiation can begin. Once differentiation occurs, the T-cells with the αβ-chain (αβT-cells) are divided into CD4 or CD8 T-cells that recognize major histocompatibility complex (MHC) class II or class I molecules, respectively. Some αβT-cells display other specific molecular residues as well as CD4 that result in regulation of the immune response. These cells are called regulatory T-cells or T_{reg}-cells. T-cells with the γδ-chain (γδT-cell) express neither CD4 nor CD8 and are relatively uncommon in the circulation but can be recruited to areas of inflammation, particularly in the gut and lungs. They interact with dendritic cells and alveolar macrophages and thereby aid in the cross talk between innate and adaptive immunity. They also seem to play a role in allergen-induced inflammation and in post-inflammation down-regulation once the inciting agent is dealt with [118].

MHC is a cell surface molecule divided into two classes and is responsible for identifying a cell as "self" or "non-self." MHC class I is located on all nucleated cells in our body whereas MHC class II is expressed by antigen-presenting cells (APCs) such as B-lymphocytes, dendritic cells and macrophages. CD8 T-cells, also called cytotoxic T-lymphocytes (CTLs), can detect if a MHC I molecule is presented to it associated with a PAMP or not and therefore either trigger tolerance to that antigen or provoke a larger immune response. MHC II requires the ingestion and processing of extracellular proteins by the APC prior to being presented to CD4 T-cells, also called helper T-cells (Th-cells) in order to initiate an immune response.

The Th-cells are further divided into three subsets called the Th-1, Th-2 and Th-17 cells based on the cytokines they synthesize once stimulated. The Th-1 cells are mainly involved in cell-mediated immunity leading to the activation of macrophages and effector T-cells such as NK-cells and CTLs. Th-2 cells are primarily involved in helping the B-cells (discussed further down) by promoting class switching and Ig production. The Th-17 cells secrete cytokines that are involved in bacterial clearance and the development of autoimmunity. Each of these T-cell subsets, once stimulated to differentiate, can lead to the up-regulation of further naïve T-cell differentiate in favor of that specific subset while down-regulating production of the others leading to a coordinated immune response.

The CTLs act slightly different. Their association with the MHC I molecule means that they are stimulated by endogenous antigens derived from intracellular pathogens. In other words, they recognize our own "self" cells when an invader gets inside and expresses a PAMP. The stimulation of the CTLs induces the production of IL-2 receptors that, when activated, lead to the cytotoxic functions of these cells. The mechanisms available to the CTLs that lead to cell death include the induction of apoptosis and production of bactericidal factors including granzymes and perforins [119].

The functions of the T-cells are mainly carried out by effector cells and are therefore part of cell-mediated immunity. The B-lymphocytes, or B-cells, act through soluble proteins called immunoglobulins (Igs), or antibodies, that can also act as B-cell receptors (BCRs) and make up part of our antibody-mediated (humoral) immunity. These can act as APCs but also develop into memory

B-cells for long-term immunity to commonly encountered antigens. There are five types of Igs (IgG, IgM, IgA, IgD and IgE) that are differentiated by their structure and function proving for a fantastic amount of variability in our immune response. Once a B-cell is stimulated to react against a specific antigen, it starts to produce antibodies directed against it and will always make antibodies to that specific antigen only. Despite this antigenic specificity, the Igs produced are not fixed. The B-cell may produce IgM initially, but can switch to making IgG or IgA, a phenomenon called class switching [120, 121]. Once the B-cell gets sensitized to an antigen, and with the help of the Th2-cells, it becomes a memory B-cell imparting lifelong reactivity against a specific antigen. When memory B-cells are activated and start to produce large amounts of antibodies, they are called plasma cells.

The cells and soluble factors that make up our immune system are vital to our body's homeostasis and essential to life. For a schematic representation of innate and adaptive immunity in the lungs and alveoli, see Figure 7.2.

7.2.2   Dendritic Cells

Dendritic cells (DCs) are unique in that they are part of our innate immune system, but they also help to bridge the gap to our adaptive immunity by acting as antigen-presenting cells. Dendritic cells and alveolar macrophages share a common progenitor as they are both derived from monocytes. Alveolar macrophages do the bulk of the work in killing or opsonizing foreign invaders but there is a point at which their capacity to do this is reached. Once this occurs, it is the dendritic cells that pick up the slack.

DCs are located throughout the alveolar interstitium, as well as several other areas in the lungs and organ systems, as an interdigitating network within the epithelium below the basement membrane. DCs are able to sample the alveolar environment due to their close proximity by sending out "tentacles" that are part of their cellular structure through the tight junctions between the alveolar epithelial cells without disrupting the integrity of this barrier [122] (see Figure 7.3). These "tentacles," called pseudopods, can "taste" the outside environment through receptors of the C-type lectin family [123]. These receptors can detect carbohydrates present on the surface of microbial organisms [124]. Once the DC encounters an antigen or particulate, they can send out cytokines and chemokines to recruit other mediators of innate immunity, such as neutrophils and macrophages, or leave the epithelium and travel to the regional lymph node where they deliver this information to naïve T-cells and provoke T-cell differentiation (see Figure 7.4). There are several subtypes of DCs present in our lungs, and each is directed towards a specific type of response. Some of these DC subtypes reside in the alveolar spaces all the time, taking samples of their environment to determine if an inflammatory response is needed. Others are recruited into areas of active infection in order to help mount a more robust immune response.

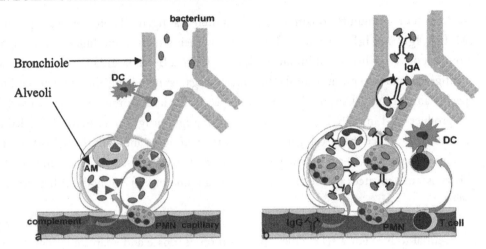

FIGURE 7.2: Schematic of the innate and adaptive immune response in the alveoli. On the left, naïve lung is exposed to bacteria; the innate defenses include the mucociliary escalator, surfactant and other secreted proteins as well as alveolar macrophages (AM) and neutrophils, or polymorphonuclear leukocytes (PMN) recruited from the blood. Inflammation facilitates the entry of complement into alveoli, where complement activation provides molecules to enhance uptake of the bacteria by phagocytes such as the alveolar macrophages. Dendritic cells (DCs) sample the bacteria, process the antigens and travel to draining lymph nodes to initiate the adaptive response. On the right, re-exposure to the bacteria leads to an amplified and more focused response. Immunoglobulin (Ig) A, secreted by plasma cells into the lumen of the conducting airways can block binding of the microbe to the epithelium. IgG in the alveolus fix themselves onto bacteria for enhanced uptake by phagocytes. Th1 cells recruited to alveoli and the interstitium, following presentation of antigen by DCs, can secrete interferon-γ to activate alveolar macrophages and recruit monocytes. (Reproduced with permission from *Respiratory Immunity to Infectious Disease*. [cited 2012 May 6]; Available from: http://www.equistro.com/download/download_files /Factsheet Respiratory immunity to infectious disease.pdf.)

The movement of DCs from their location in the alveolar interstitium to the regional lymph nodes occurs both in times of infection as well as in a normal steady state. Exactly how the DCs find their way from the alveolar interstitium to the regional lymph nodes is an area of active investigation. It seems this process involves gradients of lipid mediators (LTD$_4$, PGE$_2$ and PGD$_2$) and cytokines [124, 125, 126]. One of the most important factors involved in directing DCs to the regional lymph nodes is MMP-9 as evidenced by experiments showing that mice lacking this factor fail to mount an inflammatory reaction when exposed to an allergen [127].

DCs normally express MHC class II molecules and have low levels of co-stimulatory molecules that would otherwise provoke a full immune response. In this state, DCs are considered to

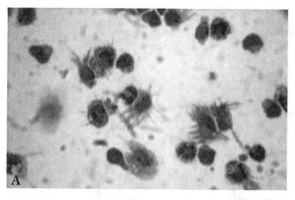 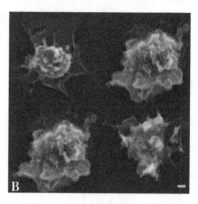

FIGURE 7.3: Pulmonary dendritic cell. Photomicrographs of pulmonary dendritic cells under light microscopy (A) and scanning electron microscopy (B). Bar is 1μm. (Reproduced with permission from Oriss, T.B., et al., *Dynamics of Dendritic Cell Phenotype and Interactions with CD4+ T Cells in Airway Inflammation and Tolerance.* The Journal of Immunology, 2005. 174(2): pp. 854–863.)

be immature, and the ingestion of harmless particulates does not change this state. When these immature DCs present to a naïve T-cell in the regional lymph node, the DC sends signals that lead to T_{reg}-cell differentiation and immunologic tolerance. An example of when this would happen is in times of normal cell turnover or non-infectious cellular injury. The affected cells undergo apoptosis and get ingested by DCs who continue to express MHC class II molecules without an associated PAMP and present this to the CD4 $\alpha\beta$T-cells. The CD4 $\alpha\beta$T-cells recognize this as a "self" antigen and initiate T_{reg}-cell formation. This type of activation is the most prominent message sent to the regional lymph nodes by DCs thereby protecting the sensitive gas exchange structures from excessive inflammation [124]. If, however, the particulate encountered by the DC is a foreign invader, the MHC II molecule on the surface of the DC will be associated with a PAMP and lead to Th-cell differentiation.

7.2.3   Other Cell Types

Dendritic cells play a very large role in the communication between the innate and adaptive immunity in our lungs, but they are not the only players. The alveolar epithelium and macrophages have a role as well. The type II pneumocytes express MHC II, and recent studies have indicated that they may be able to present antigens to Th-cells although not necessarily stimulate differentiation of naïve T-cells [128]. This is in addition to the role they play in innate immunity described above including production of antimicrobial factors and cytokines and chemokines. Alveolar macrophages, in addition to their large role in innate immunity as described earlier, can also act as APCs with

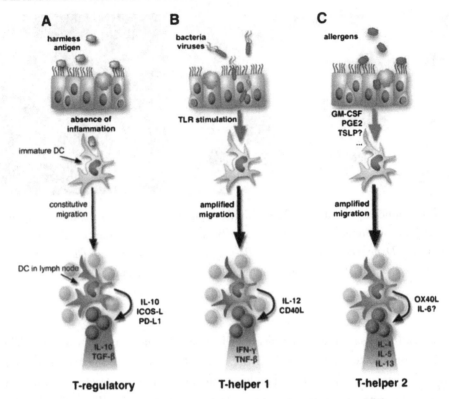

FIGURE 7.4: Dendritic cell stimulation of T-cell subsets. Schematic representation showing how DCs translate stimuli detected in the inhaled air into a specific immune response. (Reproduced with permission from Vermaelen, K. and R. Pauwels, *Pulmonary dendritic cells.* Am J Respir Crit Care Med, 2005. 172(5): pp. 530–51.)

(A) Inert antigens are continuously sampled by DCs which go through a program of migration and maturation. In the absence of inflammation (or PAMPs), expression of tolerogenic receptors and soluble mediators predominate. This leads to generation of T_{reg}-cells capable of repressing any emerging immune response to this specific antigen.

(B) Invasion by pathogens activates DCs through their TLRs. This leads to enhanced migration to the regional lymph node where antigen is presented with a PAMP leading to Th-1 cell differentiation.

(C) Allergens can also cause DC activation, albeit through poorly defined molecular signals leading to Th-2 cell differentiation.

DC, dendritic cell; IL, interleukin; ICOS-L, inducible costimulator ligand; PD-L1, programmed cell death ligand-1; TGF, tumor growth factor; TLR, Toll-like receptor; CD, cluster of differentiation; IFN, interferon, TNF, tumor necrosis factor; GM-CSF, granulocyte/monocyte-colony stimulating factor; PGE2, prostaglandin-2; TSLP, thymic stromal lymphopoietin.

MHC II molecule expression and thereby activate Th-cell differentiate and action. Recent research as shown they do this for both organic antigens and inorganic particulate matter present in polluted air [129].

7.3   ADAPTIVE IMMUNITY

Adaptive immunity, also called acquired immunity, depends on the recognition of antigens by specific T- and B-cells. These lymphocytes get "primed" to respond to antigens after repeated exposures. The more the lymphocytes are exposed to a specific antigen, the faster they can respond the next time the antigen is encountered.

The principle of adaptive immunity refers to the process of T- and B-cell differentiation described earlier. The cells and cellular products mentioned are all, at some point, located in the lungs. Whether these factors reside in the lungs waiting to be activated or are just circulating through the lungs as part of their normal life cycle depends on the cell type. The $\alpha\beta$T-cells usually circulate through and are located in the interstitial and intravascular spaces, bronchoalveolar spaces and bronchus-associated lymphoid tissues (BALT), whereas the $\gamma\delta$T-cells may reside in the intraepithelial areas [119, 130]. The process by which lymphocytes migrate from the intravascular spaces into lymph tissues, such as BALT, and then into the interstitium and alveolar spaces is a complex process involving various cytokines and adhesion factors. The way these players in adaptive immunity perform their roles in the alveolar spaces really depends on the type of response that is required.

The antibody-mediated immune response, carried out by the B-cells and their products, can be classified under three basic types. There are responses to extracellular pathogens, to autoantigens (antigens present on normal "self" cells) and those due to allergen exposure. The immune response to extracellular pathogens results from the ability of pathogens to get past the defenses of the upper airways and gain access to the alveolar spaces. Once there, they encounter some elements of humoral immunity including IgA present in the bronchial secretions but also encounter the alveolar epithelial cells and macrophages. These cells either provoke an innate immune response or lead to Th-cell stimulation via their actions as APCs. Some of the alveolar macrophages, helped by the DCs, then migrate to the regional lymph nodes where they present their antigens and MHC II molecules to naïve T-cells. This promotes T-cell differentiation and a larger immune response. Some of the invading antigens get through the blood–gas barrier and make it into the circulation where they end up in lymph nodes. These unprocessed antigens are taken up by B-cells who present them to Th2-cells who stimulate further T-cell differentiation and B-cell activation. All of these occur over about 7 to 10 days and eventually leads to the formation of memory T- and B-cells imparting a more specific and therefore quicker immune response on repeat exposure.

The immune response to autoantigens occurs when the mechanisms of immune self-tolerance fail [119]. This leads to autoreactive T- and B-cells, and their products lead to cellular injury and inappropriate death of otherwise normal, healthy tissue. Atopic, or allergic, responses occur when an immune response is generated against otherwise harmless environmental proteins. This is caused by the activation of Th2-cells and the release of IL-4 leading to B-cell activation and the production of IgE. The IL-4 released by the Th2-cells also promotes the accumulation of eosinophils and basophils leading to an inflammatory reaction. Some of the APCs that bridge the gap between innate and adaptive immunity have a role in the atopic immune response as well. The APCs in atopic individuals underproduce IL-12 and overproduce PGE_2 leading to the Th-2 cell response [131].

The T-cells themselves, as part of cell-mediated immunity, play an important role not just in the signaling of antibody production but by directly causing inflammation and promoting cytotoxic functions via the signaling of other effector cells. An example of this is the formation of granulomas. A granuloma is a specific type of inflammatory reaction involving a large collection of macrophages. They usually form in response to foreign invaders that the macrophages are not able to clear. Sometimes, the causative antigen is known, such as in *Mycobacterium tuberculosis* infection or hypersensitivity pneumonitis, while at other times, the inciting antigen is unknown, such as in sarcoidosis. This type of inflammation is a result of Th1-cell activation resulting in macrophage recruitment. Sometimes, these activated inflammatory cells collect in the interstitium of the alveoli itself resulting in alveolitis (inflammation of the alveoli) that can then lead to granuloma formation.

The CTLs have a large role to play in the management of viral infections as well as in recognition of tumor cells and in allograft (transplantation of tissues to a genetically non-identical individual) rejection. Once a foreign invader is identified and the CTL is activated, it can secrete TNF-α and IFN-γ in order to induce apoptosis or activate macrophages (among other actions), respectively. The CTLs can also release granzymes and perforins to poke holes in the membrane of the invading cells as well as several other functions. NK-cells also have cytotoxic actions that are similar to CTLs without the requirement for a MHC presentation. They are usually directed against virus-infected cells or tumor cells [117]. NK-cells can also bind IgG and facilitate antibody-mediated cell death.

. . . .

References

[1] Ochs, M., et al., The number of alveoli in the human lung. *Am J Respir Crit Care Med*, 2004. 169(1): pp. 120–4.

[2] Notter, R.H., *Lung Surfactants: Basic Science and Clinical Applications*, 2000, New York, N.Y.: Marcel Dekker.

[3] Janin, H., *The Pursuit of Learning in the Islamic World, 610-2003*, 2005, Jefferson, N.C.: McFarland. v, 229 pp.

[4] Young, J., Malpighi's "De Pulmonibus." *Proc R Soc Med*, 1929. 23(1): pp. 1–11.

[5] Sprigge, J.S., Sir Humphry Davy; his researches in respiratory physiology and his debt to Antoine Lavoisier. *Anaesthesia*, 2002. 57(4): pp. 357–64.

[6] Knight, D.M., *Humphry Davy : Science & Power. Blackwell Science Biographies*, 1992, Oxford, UK; Cambridge, USA: Blackwell. xiii, 218 pp.

[7] Severinghaus, J.W., P. Astrup, and J.F. Murray, Blood gas analysis and critical care medicine. *Am J Respir Crit Care Med*, 1998. 157(4 Pt 2): pp. S114–22.

[8] Magnus, G., Ueber die im Blute enthaltenen Gase, Sauerstoff, Stickstoff und Kohlensäure. *Annalen der Physik*, 1837. 116(4): pp. 583–606.

[9] Magnus, G., Ueber das Absorptionsvermögen des Bluts für Sauerstoff. *Annalen der Physik*, 1845. 142(10): pp. 177–206.

[10] Lavoisier, A.L., France. Ministère de l'éducation nationale., and É. Grimaux, *Œuvres de Lavoisier*, 1862, Paris: Imprimerie impériale.

[11] Joseph Black—rediscoverer of fixed air. *JAMA*, 1966. 196(4): pp. 362–3.

[12] Voelkel, N.F. and W. MacNee, *Chronic Obstructive Lung Diseases*, 2002, Hamilton, Ont.: BC Decker. xi, 428 pp.

[13] Boyden, E.A., Development and growth of the airways, in *Development of the Lung*, W.A. Hudson, Editor. 1977, M. Dekker: New York. pp. 30–5.

[14] The Free Dictionary. *The American Heritage Dictionary of the English Language 2000*, 2009 [cited 2012 May 6]; 4th:[Available from: http://www.thefreedictionary.com/Monopodial.

[15] Dichotomous branching. *Encyclopedia Britannica*, 2012 [cited 2012 May 6]; Available from: http://www.britannica.com/EBchecked/topic/162091/dichotomous-branching.

[16]   Shannon, J.M. and J.M. Greenberg, Lung growth and development, in *Murray and Nadel's Textbook of Respiratory Medicine*, J.F. Murray and R.J. Mason, Editors. 2010, Saunders/Elsevier: Philadelphia, PA. pp. 26–37.

[17]   Shannon, J.M. and B.A. Hyatt, Epithelial-mesenchymal interactions in the developing lung. *Annu Rev Physiol*, 2004. 66: pp. 625–45.

[18]   Merkus, P.J., A.A. ten Have-Opbroek, and P.H. Quanjer, Human lung growth: a review. *Pediatr Pulmonol*, 1996. 21(6): pp. 383–97.

[19]   Potter, E.L. and C.G. Loosli, Prenatal development of the human lung. *AMA Am J Dis Child*, 1951. 82(2): pp. 226–8.

[20]   Burri, P.H., et al., Lung development, in *Clinical Physiology Series*, 1999, Published for the American Physiological Society by Oxford University Press: New York. p. xiii, 451 pp.

[21]   Prodhan, P. and T.B. Kinane, Developmental paradigms in terminal lung development. *Bioessays*, 2002. 24(11): pp. 1052–9.

[22]   Burri, P.H. and M. Moschopulos, Structural analysis of fetal rat lung development. *Anat Rec*, 1992. 234(3): pp. 399–418.

[23]   Sauve, R.S., et al., Before viability: a geographically based outcome study of infants weighing 500 grams or less at birth. *Pediatrics*, 1998. 101(3 Pt 1): pp. 438–45.

[24]   Costeloe, K., et al., The EPICure study: outcomes to discharge from hospital for infants born at the threshold of viability. *Pediatrics*, 2000. 106(4): pp. 659–71.

[25]   Wendel, D.P., et al., Impaired distal airway development in mice lacking elastin. *Am J Respir Cell Mol Biol*, 2000. 23(3): pp. 320–6.

[26]   Schittny, J.C., et al., Programmed cell death contributes to postnatal lung development. *Am J Respir Cell Mol Biol*, 1998. 18(6): pp. 786–93.

[27]   Burri, P.H., Structural aspects of postnatal lung development—alveolar formation and growth. *Biol Neonate*, 2006. 89(4): pp. 313–22.

[28]   Burri, P.H., The postnatal growth of the rat lung. 3. Morphology. *Anat Rec*, 1974. 180(1): pp. 77–98.

[29]   Dryden, R. Respiratory system. *Bionalogy: Learning Healthcare Biology through Analogy 2004*, November 6, 2011 [cited 2012 April 8]; Available from: http://www.bionalogy.com /respiratory_system.htm.

[30]   Scarpelli, E., Perinatal lung mechanics and the first breath. *Lung*, 1984. 162(1): pp. 61–71.

[31]   Macklem, P.T., Airway obstruction and collateral ventilation. *Physiol Rev*, 1971. 51(2): pp. 368–436.

[32]   Bastacky, J. and J. Goerke, Pores of Kohn are filled in normal lungs: low-temperature scanning electron microscopy. *J Appl Physiol*, 1992. 73(1): pp. 88–95.

[33]   Spencer, H. and D. Leof, The innervation of the human lung. *J Anat*, 1964. 98: pp. 599–609.

[34]   Larsell, G. and R. Dow, The innervation of the human lung. *Am J Anat*, 1933. 52: p. 125.

[35]   Paintal, A.S., Mechanism of stimulation of type J pulmonary receptors. *J Physiol*, 1969. 203(3): pp. 511–32.

[36]   Lauweryns, J.M. and P. Goddeeris, Neuroepithelial bodies in the human child and adult lung. *Am Rev Respir Dis*, 1975. 111(4): pp. 469–76.

[37]   Said, S.I. and V. Mutt, Long acting vasodilator peptide from lung tissue. *Nature*, 1969. 224: pp. 699–700.

[38]   Lauweryns, J.M., A.T. Van Lommel, and R.J. Dom, Innervation of rabbit intrapulmonary neuroepithelial bodies. Quantitative and qualitative ultrastructural study after vagotomy. *J Neurol Sci*, 1985. 67(1): pp. 81–92.

[39]   Crapo, J.D., et al., Cell number and cell characteristics of the normal human lung. *Am Rev Respir Dis*, 1982. 125(6): pp. 740–5.

[40]   Gonzalez, R.F., L. Allen, and L.G. Dobbs, Rat alveolar type I cells proliferate, express OCT-4, and exhibit phenotypic plasticity in vitro. *Am J Physiol Lung Cell Mol Physiol*, 2009. 297(6): pp. L1045–55.

[41]   Uhal, B.D., Cell cycle kinetics in the alveolar epithelium. *Am J Physiol*, 1997. 272(6 Pt 1): pp. L1031–45.

[42]   Evans, M.J. and J.D. Hackney, Cell proliferation in lungs of mice exposed to elevated concentrations of oxygen. *Aerosp Med*, 1972. 43(6): pp. 620–2.

[43]   Farrell, P.M. and M.E. Avery, Hyaline membrane disease. *Am Rev Respir Dis*, 1975. 111(5): pp. 657–88.

[44]   Trapnell, B.C., J.A. Whitsett, and K. Nakata, Pulmonary alveolar proteinosis. *N Engl J Med*, 2003. 349(26): pp. 2527–39.

[45]   Melton, K.R., et al., SP-B deficiency causes respiratory failure in adult mice. *Am J Physiol Lung Cell Mol Physiol*, 2003. 285(3): pp. L543–9.

[46]   Williams, G.D., et al., Surfactant protein B deficiency: clinical, histological and molecular evaluation. *J Paediatr Child Health*, 1999. 35(2): pp. 214–20.

[47]   Glasser, S.W., et al., Altered stability of pulmonary surfactant in SP-C-deficient mice. *Proc Natl Acad Sci U S A*, 2001. 98(11): pp. 6366–71.

[48]   Nogee, L.M., et al., A mutation in the surfactant protein C gene associated with familial interstitial lung disease. *N Engl J Med*, 2001. 344(8): pp. 573–9.

[49]   Kishore, U., et al., Surfactant proteins SP-A and SP-D: structure, function and receptors. *Mol Immunol*, 2006. 43(9): pp. 1293–315.

[50]   Kuroki, Y., M. Takahashi, and C. Nishitani, Pulmonary collectins in innate immunity of the lung. *Cell Microbiol*, 2007. 9(8): pp. 1871–9.

[51]   Panos, R.J., et al., Keratinocyte growth factor and hepatocyte growth factor/scatter factor

are heparin-binding growth factors for alveolar type II cells in fibroblast-conditioned medium. *J Clin Invest*, 1993. 92(2): pp. 969–77.

[52] Morikawa, O., et al., Effect of adenovector-mediated gene transfer of keratinocyte growth factor on the proliferation of alveolar type II cells in vitro and in vivo. *Am J Respir Cell Mol Biol*, 2000. 23(5): pp. 626–35.

[53] Danto, S.I., et al., Reversible transdifferentiation of alveolar epithelial cells. *Am J Respir Cell Mol Biol*, 1995. 12(5): pp. 497–502.

[54] Kim, K.K., et al., Alveolar epithelial cell mesenchymal transition develops in vivo during pulmonary fibrosis and is regulated by the extracellular matrix. *Proc Natl Acad Sci U S A*, 2006. 103(35): pp. 13180–5.

[55] Kajstura, J., et al., Evidence for human lung stem cells. *N Engl J Med*, 2011. 364(19): pp. 1795–806.

[56] Crapo, J.D., et al., Cell number and cell characteristics of the normal human lung. *Am Rev Respir Dis*, 1982. 126(2): pp. 332–7.

[57] Herzog, E.L., et al., Knowns and unknowns of the alveolus. *Proc Am Thorac Soc*, 2008. 5(7): pp. 778–82.

[58] Brody, J.S. and N.B. Kaplan, Proliferation of alveolar interstitial cells during postnatal lung growth. Evidence for two distinct populations of pulmonary fibroblasts. *Am Rev Respir Dis*, 1983. 127(6): pp. 763–70.

[59] Hashimoto, N., et al., Bone marrow-derived progenitor cells in pulmonary fibrosis. *J Clin Invest*, 2004. 113(2): pp. 243–52.

[60] Phillips, R.J., et al., Circulating fibrocytes traffic to the lungs in response to CXCL12 and mediate fibrosis. *J Clin Invest*, 2004. 114(3): pp. 438–46.

[61] Alexander, J.S. and J.W. Elrod, Extracellular matrix, junctional integrity and matrix metalloproteinase interactions in endothelial permeability regulation. *J Anat*, 2002. 200(6): pp. 561–74.

[62] Stevens, T., et al., Lung vascular cell heterogeneity: endothelium, smooth muscle, and fibroblasts. *Proc Am Thorac Soc*, 2008. 5(7): pp. 783–91.

[63] Parker, J.C. and S. Yoshikawa, Vascular segmental permeabilities at high peak inflation pressure in isolated rat lungs. *Am J Physiol Lung Cell Mol Physiol*, 2002. 283(6): pp. L1203–9.

[64] King, J., et al., Structural and functional characteristics of lung macro- and microvascular endothelial cell phenotypes. *Microvasc Res*, 2004. 67(2): pp. 139–51.

[65] Cioffi, D.L., et al., Dominant regulation of interendothelial cell gap formation by calcium-inhibited type 6 adenylyl cyclase. *J Cell Biol*, 2002. 157(7): pp. 1267–78.

[66] Grishko, V., et al., Oxygen radical-induced mitochondrial DNA damage and repair in pul-

monary vascular endothelial cell phenotypes. *Am J Physiol Lung Cell Mol Physiol*, 2001. 280(6): pp. L1300–8.

[67]   Guyton, A.C., *Textbook of Medical Physiology, 9th ed* 1996, Philadelphia: W.B. Saunders.

[68]   Studdy, P.R., R. Lapworth, and R. Bird, Angiotensin-converting enzyme and its clinical significance—a review. *J Clin Pathol*, 1983. 36(8): pp. 938–47.

[69]   Sherman, T.S., et al., Nitric oxide synthase isoform expression in the developing lung epithelium. *Am J Physiol*, 1999. 276(2 Pt 1): pp. L383–90.

[70]   Wiener-Kronish, J.P., K.H. Albertine, and M.A. Matthay, Differential responses of the endothelial and epithelial barriers of the lung in sheep to *Escherichia coli* endotoxin. *J Clin Invest*, 1991. 88(3): pp. 864–75.

[71]   Albert, R.K., et al., Increased surface tension favors pulmonary edema formation in anesthetized dogs' lungs. *J Clin Invest*, 1979. 63(5): pp. 1015–8.

[72]   Taylor, A.E. and K.A. Gaar, Jr., Estimation of equivalent pore radii of pulmonary capillary and alveolar membranes. *Am J Physiol*, 1970. 218(4): pp. 1133–40.

[73]   Gorin, A.B. and P.A. Stewart, Differential permeability of endothelial and epithelial barriers to albumin flux. *J Appl Physiol*, 1979. 47(6): pp. 1315–24.

[74]   Taylor, A.E. and J.C. Parker, in *Handbook of Physiology. Section 3: The Respiratory System. Vol 1: Circulation and Nonrespiratory Function*, A. Fishman and A. Fisher, Editors. 1985, American Physiological Society: Bethsda, MD. pp. 167–230.

[75]   Kozono, D., et al., Aquaporin water channels: atomic structure molecular dynamics meet clinical medicine. *J Clin Invest*, 2002. 109(11): pp. 1395–9.

[76]   Zhang, P., et al., Innate immunity and pulmonary host defense. *Immunol Rev*, 2000. 173: pp. 39–51.

[77]   Stahlhofen, W., J. Gebhart, and J. Heyder, Experimental determination of the regional deposition of aerosol particles in the human respiratory tract. *Am Ind Hyg Assoc J*, 1980. 41(6): pp. 385–98a.

[78]   Stuart, B.O., Deposition and clearance of inhaled particles. *Environ Health Perspect*, 1984. 55: pp. 369–90.

[79]   Green, G.M., The J. Burns Amberson Lecture—in defense of the lung. *Am Rev Respir Dis*, 1970. 102(5): pp. 691–703.

[80]   Murray, J.F., *The Normal Lung* 1976, Philadelphia, PA: W.B. Saunders.

[81]   Mason, R.J., Biology of alveolar type II cells. *Respirology*, 2006. 11: pp. S12–5.

[82]   Martin, T.R. and C.W. Frevert, Innate immunity in the lungs. *Proc Am Thorac Soc*, 2005. 2(5): pp. 403–11.

[83]   Gardai, S.J., et al., By binding SIRPalpha or calreticulin/CD91, lung collectins act as dual

function surveillance molecules to suppress or enhance inflammation. *Cell*, 2003. 115(1): pp. 13–23.

[84] LeVine, A.M. and J.A. Whitsett, Pulmonary collectins and innate host defense of the lung. *Microbes Infect*, 2001. 3(2): pp. 161–6.

[85] Wong, H.R., et al., Induction of lipopolysaccharide-binding protein gene expression in cultured rat pulmonary artery smooth muscle cells by interleukin 1 beta. *Am J Respir Cell Mol Biol*, 1995. 12(4): pp. 449–54.

[86] Dentener, M.A., et al., Production of the acute-phase protein lipopolysaccharide-binding protein by respiratory type II epithelial cells: implications for local defense to bacterial endotoxins. *Am J Respir Cell Mol Biol*, 2000. 23(2): pp. 146–53.

[87] Skerrett, S.J., et al., Respiratory epithelial cells regulate lung inflammation in response to inhaled endotoxin. *Am J Physiol Lung Cell Mol Physiol*, 2004. 287(1): pp. L143–52.

[88] Riches, D. and M. Fenton, *Monocytes, Macrophages, and Dendritic Cells of the Lung*, in *Murray and Nadel's textbook of respiratory medicine* 2005, Saunders: Philadelphia, PA. pp. 355–76.

[89] Harmsen, A.G., et al., The role of macrophages in particle translocation from lungs to lymph nodes. *Science*, 1985. 230(4731): pp. 1277–80.

[90] Doherty, D.E., et al., Prolonged monocyte accumulation in the lung during bleomycin-induced pulmonary fibrosis. A noninvasive assessment of monocyte kinetics by scintigraphy. *Lab Invest*, 1992. 66(2): pp. 231–42.

[91] Albert, D.H. and F. Snyder, Biosynthesis of 1-alkyl-2-acetyl-sn-glycero-3-phosphocholine (platelet-activating factor) from 1-alkyl-2-acyl-sn-glycero-3-phosphocholine by rat alveolar macrophages. Phospholipase A2 and acetyltransferase activities during phagocytosis and ionophore stimulation. *J Biol Chem*, 1983. 258(1): pp. 97–102.

[92] Hsueh, W., Prostaglandin biosynthesis in pulmonary macrophages. *Am J Pathol*, 1979. 97(1): pp. 137–48.

[93] Martin, T.R., et al., Relative contribution of leukotriene B4 to the neutrophil chemotactic activity produced by the resident human alveolar macrophage. *J Clin Invest*, 1987. 80(4): pp. 1114–24.

[94] Lohmann-Matthes, M., C. Steinmuller, and G. Franke-Ullmann, Pulmonary macrophages. *Eur Respir J*, 1994. 7(9): pp. 1678–89.

[95] Standiford, T.J., et al., Interleukin-8 Gene Expression from Human Alveolar Macrophages: The Role of Adherence. *Am J Respir Cell Mol Biol*, 1991. 5(6): pp. 579–85.

[96] Driscoll, K.E., et al., Macrophage inflammatory proteins 1 and 2: expression by rat alveolar macrophages, fibroblasts, and epithelial cells and in rat lung after mineral dust exposure. *Am J Respir Cell Mol Biol*, 1993. 8(3): pp. 311–8.

[97] Martinet, Y., et al., Exaggerated spontaneous release of platelet-derived growth factor by

alveolar macrophages from patients with idiopathic pulmonary fibrosis. *N Engl J Med*, 1987. 317(4): pp. 202–9.

[98]   Jordana, M., et al., Spontaneous in vitro release of alveolar-macrophage cytokines after the intratracheal instillation of bleomycin in rats. Characterization and kinetic studies. *Am Rev Respir Dis*, 1988. 137(5): pp. 1135–40.

[99]   Bitterman, P.B., et al., Human alveolar macrophage growth factor for fibroblasts. Regulation and partial characterization. *J Clin Invest*, 1982. 70(4): pp. 806–22.

[100] Bitterman, P.B., et al., Role of fibronectin as a growth factor for fibroblasts. *J Cell Biol*, 1983. 97(6): pp. 1925–32.

[101] Kunkel, S.L., et al., The role of chemokines in the immunopathology of pulmonary disease. *Forum (Genova)*, 1999. 9(4): pp. 339–55.

[102] Riches, D., et al., Innate imunity in the lungs, in *Murray and Nadel's Textbook of Respiratory Medicine*, J.F. Murray and R.J. Mason, Editors. 2010, Saunders/Elsevier: Philadelphia, PA.

[103] Morris, D.G., et al., Loss of integrin alpha(v)beta6-mediated TGF-beta activation causes Mmp12-dependent emphysema. *Nature*, 2003. 422(6928): pp. 169–73.

[104] Munger, J.S., et al., The integrin alpha v beta 6 binds and activates latent TGF beta 1: a mechanism for regulating pulmonary inflammation and fibrosis. *Cell*, 1999. 96(3): pp. 319–28.

[105] Kawabe, T., et al., Immunosuppressive activity induced by nitric oxide in culture supernatant of activated rat alveolar macrophages. *Immunology*, 1992. 76(1): pp. 72–8.

[106] Roth, M.D. and S.H. Golub, Human pulmonary macrophages utilize prostaglandins and transforming growth factor beta 1 to suppress lymphocyte activation. *J Leukoc Biol*, 1993. 53(4): pp. 366–71.

[107] Nicod, L.P., Lung defences: an overview. *European Respiratory Review*, 2005. 14(95): pp. 45–50.

[108] Smith, D.L. and R.D. Deshazo, Bronchoalveolar Lavage in Asthma. *Am Rev Respir Dis*, 1993. 148: pp. 523–32.

[109] Zarbock, A. and K. Ley, Neutrophil adhesion and activation under flow. *Microcirculation*, 2009. 16(1): pp. 31–42.

[110] Jones, H.A., et al., In vivo measurement of neutrophil activity in experimental lung inflammation. *Am J Respir Crit Care Med*, 1994. 149(6): pp. 1635–9.

[111] Hidalgo, A., et al., Heterotypic interactions enabled by polarized neutrophil microdomains mediate thromboinflammatory injury. *Nat Med*, 2009. 15(4): pp. 384–91.

[112] Brinkmann, V., et al., Neutrophil Extracellular Traps Kill Bacteria. *Science*, 2004. 303(5663): pp. 1532–5.

[113] von Kockritz-Blickwede, M. and V. Nizet, Innate immunity turned inside-out: antimicrobial defense by phagocyte extracellular traps. *J Mol Med (Berl)*, 2009. 87(8): pp. 775–83.

[114] Jin, F.Y., et al., Secretory leukocyte protease inhibitor: a macrophage product induced by and antagonistic to bacterial lipopolysaccharide. *Cell*, 1997. 88(3): pp. 417–26.

[115] Serhan, C.N., N. Chiang, and T.E. Van Dyke, Resolving inflammation: dual anti-inflammatory and pro-resolution lipid mediators. *Nat Rev Immunol*, 2008. 8(5): pp. 349–61.

[116] Skerrett, S.J., et al., Role of the type 1 TNF receptor in lung inflammation after inhalation of endotoxin or *Pseudomonas aeruginosa*. *Am J Physiol*, 1999. 276(5 Pt 1): pp. L715–27.

[117] Yokoyama, W.M., S. Kim, and A.R. French, The dynamic life of natural killer cells. *Annu Rev Immunol*, 2004. 22: pp. 405–29.

[118] Born, W.K., et al., Role of Gamma/Delta T Cells in Lung Inflammation. *Open Immunol J*, 2009. 2: pp. 143–150.

[119] Fontenot, A.P. and P.L. Simonian, Adaptive immunity, in *Murray and Nadel's Textbook of Respiratory Medicine*, J.F. Murray and R.J. Mason, Editors. 2010, Saunders/Elsevier: Philadelphia, PA.

[120] Torres, R., J. Imboden, and H.J. Schroeder, Antigen receptor genes, gene products and co-receptors, in *Clinical Immunology: Principles and Practice*, R. Rich, et al., Editors. 2008, Mosby: London. pp. 53–78.

[121] Stavnezer, J., J.E. Guikema, and C.E. Schrader, Mechanism and regulation of class switch recombination. *Annu Rev Immunol*, 2008. 26: pp. 261–92.

[122] Rescigno, M., et al., Dendritic cells express tight junction proteins and penetrate gut epithelial monolayers to sample bacteria. *Nat Immunol*, 2001. 2(4): pp. 361–7.

[123] Engering, A., T.B. Geijtenbeek, and Y. van Kooyk, Immune escape through C-type lectins on dendritic cells. *Trends Immunol*, 2002. 23(10): pp. 480–5.

[124] Vermaelen, K. and R. Pauwels, Pulmonary dendritic cells. *Am J Respir Crit Care Med*, 2005. 172(5): pp. 530–51.

[125] Sozzani, S., et al., Differential regulation of chemokine receptors during dendritic cell maturation: a model for their trafficking properties. *J Immunol*, 1998. 161(3): pp. 1083–6.

[126] Yanagihara, S., et al., EBI1/CCR7 is a new member of dendritic cell chemokine receptor that is up-regulated upon maturation. *J Immunol*, 1998. 161(6): pp. 3096–102.

[127] Vermaelen, K.Y., et al., Matrix metalloproteinase-9-mediated dendritic cell recruitment into the airways is a critical step in a mouse model of asthma. *J Immunol*, 2003. 171(2): pp. 1016–22.

[128] Debbabi, H., et al., Primary type II alveolar epithelial cells present microbial antigens to antigen-specific CD4+ T cells. *Am J Physiol - Lung Cell Mol Physiol*, 2005. 289(2): pp. L274–9.

[129] Miyata, R. and S.F. van Eeden, The innate and adaptive immune response induced by alveolar macrophages exposed to ambient particulate matter. *Toxicol Appl Pharmacol*, 2011. 257(2): pp. 209–26.

[130] Born, W.K., C.L. Reardon, and R.L. O'Brien, The function of gammadelta T cells in innate immunity. *Curr Opin Immunol*, 2006. 18(1): pp. 31–8.

[131] van der Pouw Kraan, T.C., et al., Reduced production of IL-12 and IL-12-dependent IFN-gamma release in patients with allergic asthma. *J Immunol*, 1997. 158(11): pp. 5560–5.

[132] Phases of lung development. *Module 18: Respiration Tract* [cited 2012 March 18]; Available from: http://www.embryology.ch/anglais/rrespiratory/phasen01.html.

Author Biographies

Dr. D. Keith Payne is the Bryn Professor of Medicine in the Division of Pulmonary, Critical Care, and Sleep Medicine at Louisiana State University Health Sciences Center in Shreveport. He completed his undergraduate studies at Washington and Lee University in Lexington, Virginia, and received his M.D. degree from the University of Texas Medical Branch in Galveston, Texas. He completed his training in Internal Medicine and fellowship in Pulmonary/Critical Care medicine at LSU Health Sciences Center. His research interests have focused on lung microvascular permeability changes in response to oxidant injury and from tumor cell/endothelial cell interactions. He is actively involved in the teaching programs for medical students, residents, and subspecialty fellows in pulmonary/critical care and attends on the pulmonary consultation services at LSU and the outpatient clinics.

Dr. Adam Wellikoff is an Assistant Professor of Medicine in the Division of Pulmonary, Critical Care, and Sleep Medicine at Louisiana State University Health Sciences Center in Shreveport. He completed his undergraduate studies at the University of Florida in Gainesville, Florida, and received his M.D. degree from Saba University School of Medicine in the Netherland-Antilles. He completed his Internal Medicine training at Flushing Hospital Medical Center in Queens, New York, and his fellowship in Pulmonary and Critical Care Medicine at LSU Health Sciences Center. His focus is primarily in Interventional Pulmonology, and he is actively involved in the teaching program at LSU for medical students, residents, and fellows. Dr. Wellikoff's research activities are directed toward pleural cryotherapy and confocal laser endomicroscopic imaging of the structure of the peripheral areas of the lungs.

Colloquium Series on Integrated Systems Physiology: From Molecule to Function to Disease

Respiratory

Respiratory Muscles: Structure, Function & Regulation
Heather Gransee and Gary Sieck
Mayo Clinic

Airway Epithelium
Jonathan Widdicombe
UC-Davis

Cardiovascular

Angiogenesis
Thomas Adair and J.P. Montani

Capillary Fluid Exchange: Regulation, Functions, and Pathology
Joshua Scallan, Virginia Huxley, Ronald Korthuis
University of Missouri-Columbia

Cardiovascular Responses to Exercise
Lusha Xiang and Robert Hester
LSU Health Sciences, Shreveport

The Cerebral Circulation
Marilyn Cipolla
University of Vermont

Control of Cardiac Output
David Young
U. Mississippi Medical Center

The Endothelium, Part I: Multiple Functions of the Endothelial Cells—Focus on Endothelium-Derived Vasoactive Mediators
Michel Félétou
Institut de RecherchesServier, France

The Endothelium, Part II: EDHF-Mediated Responses "The Classical Pathway"
Michel Félétou
Institut de RecherchesServier, France

Hemorheology & Hemodynamics
Giles R. Cokelet
Montana State University, *emeritus*

Hepatic Circulation
Wayne Lautt
University of Manitoba, Canada

Inflammation and the Microcirculation
D. Neil Granger and Elena Senchenkova

Local Control of Microvascular Perfusion
Michael Hill and Michael J. Davis
University of Missouri-Columbia

The Ocular Circulation
Jeffrey Kiel
U. Texas Health Sciences

Platelet-Vessel Wall Interactions in Hemostasis and Thrombosis
Rolando Rumbaut and PerumalThiagarajan
Baylor College of Medicine

Reactive Oxygen Species and the Cardiovascular System
RhianTouyz and Augusto C. Montezano
University of Glasgow, Scotland

Regulation of Cardiac Contractility
R. John Solaro
University of Illinois-Chicago

Regulation of Endothelial Barrier Function
Sarah Yuan and Robert R. Rigor
UC-Davis

Regulation of Tissue Oxygenation.
Roland Pittman
Virginia Commonwealth University

Regulation of Vascular Smooth Muscle Function
Raouf Khalil
Harvard Medical School

Skeletal Muscle Circulation
Ronald Korthuis,
University of Missouri-Columbia

Vascular Biology of the Placenta
Yuping Wang
Louisiana State University

Gastrointestinal

Colonic Motility: From Bench Side to Bedside
Sushil K. Sarna

Enteric Nervous System: The Brain-in-the-Gut
Jackie D. Wood
Ohio State University

Intestinal Immune System
Soichiro Miura, RyotaHokari, ShunsukeKomoto
National Defense Medical College, Japan

Intestinal Water and Electrolyte Transport in Health and Disease
Mrinalini C. Rao, JayashreeSarathy (nee Venkatasubramanian), Mei Ao
University of Illinois-Chicago

Motor Function of the Pharynx, Esophagus & Its Sphincters
Ravinder K. Mittal
UC-San Diego

Neural Control of Gastrointestinal Function
David Grundy and Simon Brookes
University of Sheffield (UK) and Flinders University (AU)

Regulation of Gastrointestinal Mucosal Growth
J.Y. Wang and Rao N. Jaladanki
University of Maryland

The Biliary System
David H.Q. Wang, Brent A. Neuschwander-Tetri, and PieroPortincasa
St. Louis University

The Enteric Microbiota
Francisco Guarner
University Hospital Valld'Hebron, Barcelona

The Exocrine Pancreas
Stephen J. Pandol
UCLA

The Gastrointestinal Circulation
Peter Kvietys
Alfaisal University, Saudi Arabia

Renal

Endothelin in the Kidney
David Pollock and Erika Boesen
Georgia Health Sciences

HemeOxygenase and the Kidney
David E. Stec
University of Mississippi Medical Center

Neural Control of Renal Function
Ulla Kopp
University of Iowa